GAS ENGINES AND CO-GENERATION

GAS ENGINES AND CO-GENERATION

Papers presented at a Seminar organized by the Combustion Engines Group of the Institution of Mechanical Engineers, and held at the National Motor Museum, Solihull, West Midlands, on 10/11 May 1990.

Published by
Mechanical Engineering Publications Limited for
The Institution of Mechanical Engineers
LONDON

First Published 1990

© The Institution of Mechanical Engineers 1990

ISBN 0 85298 732 3

A CIP catalogue record for this book is available from the British Library.

Printed by Waveney Print Services Ltd, Beccles, Suffolk

CONTENTS

Proceedings of the Institution of Mechanical Engineers

The *Proceedings* is, and always has been, the premier publication of the IMechE. Since the inception of the Institution in 1847 the *Proceedings* has endeavoured to satisfy the requirements of its audience. The publication has progressed from being a random collection of separate papers dealing with matters of Institutional as well as engineering interest, to being a multi-part learned society serial. This development has culminated in an eight-part, twenty-eight issue publication.

Each Part has its own Editorial Panel which, together with a number of international corresponding members, is guided by an eminent engineer acting as Editor. Each and every paper published in the *Proceedings* must meet the highest possible standards of excellence.

ISSN 0020/3483 The Set

1990 Subscription Rates

Part A · **Journal of Power and Energy**
Editor: Professor A J Reynolds, Brunel University, Uxbridge, England
Published quarterly starting in January   ISSN 0957/6509   UK **£87.00**   Export **£109.00**

Part B · **Journal of Engineering Manufacture**
Editor: Professor A N Bramley, University of Bath, Bath, England
Published quarterly starting in February   ISSN 0954/4054   UK **£87.00**   Export **£109.00**

Part C · **Journal of Mechanical Engineering Science**
Editor: Professor R C Baker, Cranfield Institute of Technology, Cranfield, Bedford, England
Published six times a year starting in February   ISSN 0954/4062   UK **£114.00**   Export **£142.00**

Part D · **Journal of Automobile Engineering**
Editor: Mr M Lewis, formerly Gaydon Technology Limited, Lighthorne Heath, Warwickshire, England
Published quarterly starting in March   ISSN 0954/4070   UK **£87.00**   Export **£109.00**

Part E · **Journal of Process Mechanical Engineering**
Editor: Dr G Thompson, UMIST, Manchester, England
Published twice a year in January and July   ISSN 0954/4089   UK **£50.00**   Export **£63.00**

Part F · **Journal of Rail and Rapid Transit**
Editor: Mr D J W Souch, formerly GEC Transportation Projects, Manchester, England
Published twice a year in March and September   ISSN 0954/4097   UK **£50.00**   Export **£63.00**

Part G · **Journal of Aerospace Engineering**
Editor: Mr F K E Behennah, formerly Flight Refuelling Limited, Wimborne, Dorset, England
Published twice a year in May and November   ISSN 0954/4100   UK **£50.00**   Export **£63.00**

Part H · **Journal of Engineering in Medicine**
Editor: Professor D Dowson, University of Leeds, Leeds, England
Published quarterly starting in February   ISSN 0954/4119   UK **£87.00**   Export **£109.00**

SAVE by taking a SPECIAL COMBINED RATE

Combined Price Parts A – G   Save 20%   UK **£423.00**   Export **£528.00**

Combined Price Parts A – H   Save 25%   UK **£462.00**   Export **£578.00**

All prices include postage and packing.

Copies sent outside the UK benefit from air-speeded despatch to ensure fast delivery.

Sample copies available on request.

Orders and enquiries to: Sales Department, Mechanical Engineering Publications Limited, P O Box 24, Northgate Avenue, Bury St Edmunds, Suffolk, IP32 6BW, England

The Beginner's Guide to CHP

I S STARK, BSc(Eng), CEng, MIMechE
Drummond, Stark & Partners, Consulting Engineers, UK

INTRODUCTION

1.   This paper is specifically for those who are new to the subject of C.H.P.   The aim is to outline the basics, explain some of the terminology so that you can more easily understand the papers to follow and, hopefully, leave you with enough information to find out where you can learn more later.

2.   It is amazing how many people, including Architects & Consulting Engineers involved in large projects such as building Hospitals and Hotels confess to knowing little about this subject.   There is also some prejudice against C.H.P. - mainly due to poor design, reliability and performance of small scale C.H.P. in the earlier days of the industry.

3.   Combined Heat and Power in the form of District Heating Systems has been around in Great Britain for some time - again having a chequered career.   One example of success, however, was Battersea Power Station built in 1953, completed in 1956 and dismantled in the early '80's.   This was called the Pimlico Scheme where hot water, supplied from back pressure turbines, was piped underneath the river Thames to supply 2,403 flats on the North Embankment in an area of 31 acres. The district heating system still survives but, sadly, the heat is now generated by dual-fuelled boilers.   Recently a leisure complex was added to the scheme - a good opportunity to incorporate a C.H.P. unit - but C.H.P., although considered, was rejected.

4.   Small-Scale C.H.P. (up to .5 MU) was given a boost in this country by the Energy Act of 1983 which allowed Private Producers of Electricity to:-

   a. export excess electricity to the Grid.
   b. export electricity to a specified destination or customer.

Whether this can be done profitably, or at a reasonable price, after privatisation of electricity remains to be seen.

WHAT IS C.H.P.?

5.   It may help to explain the main components of a Combined Heat and Power System by referring to the line drawing at Figure 1.

6.   Approximately 25% of the total fuel input is used to generate electricity from the engine which, in to-day's context, is a spark-ignition reciprocating gas engine.
The generator can be either:-

   a. Synchronous, which is self-excited and essential to any stand-alone system which can be used for stand-by generation of electricity

   OR

   b. Asynchronous, which is Mains excited - simpler, cheaper and more efficient - but not suitable for stand-by use.

7.   The electricity generated must be controlled and monitored.   The minimum requirement being governed by Electricity Council Engineering Regulation G59 which are the 'Recommendations for the connection of Private Generating Plants to the Electricity Board's Distribution System'.

8.   Of the remaining 75% of the energy supplied to the engine about 65% of it can be recovered from the engine jacket coolant and exhaust gases via heat exchangers.   This is where C.H.P. scores over the conventional power station which generally dissipates this 'waste' heat via vast cooling towers.

9.   Preferably, the engine should be Industrial - i.e.

   a. designed and developed to provide reliable power

   b. large components to retain heat

   c. large bearing areas and low speeds for minimum wear

   d. built for ease of maintenance

1

As always, a good maintenance programme is essential to keep reliability up and costs down. Most suppliers of C.H.P. systems offer sophisticated service and parts contracts - sometimes including centralised, computer-controlled monitoring systems using Telemetry. It is worth adding a word of caution here about the supply of Gas. British Gas normally supply gas at between 8" and 10" of water gauge. This should be sufficient to run a gas engine but the legal minimum for supply is 5½" of water gauge - well below the optimum level. The reason for British Gas wanting to supply at as low a pressure as possible is not only for safety reasons but also it keeps the rate of losses to a minimum.

10. The heat equation for C.H.P. is illustrated in the SANKEY DIAGRAM in Figure 2 which is the Energy Industry's standard for showing the energy flows of a process or site. This can be simple, as shown, or complex according to the information available

11. MONEY

Money is one of the most important aspects of C.H.P. Much as we might appeal to commerce to save energy and use it so that our planet is not depleted too quickly of its resources, C.H.P. must be seen as a money-earner and saver if it is to prosper.

12. Both the Electricity and Gas Boards publish Tariffs. These are easily obtainable from showrooms but understanding them is a different matter.
The Electricity Tariff is so complex that a whole industry has grown up in claiming the best rate.
Below are some average figures for prices:-

   a. Cost of Electricity
      from the Board   = 4.5p/k.Whs

   b. Cost of Electricity
      to generate from C.H.P.   = 2.0p/k.Whs

   c. Cost of Maintenance of
      small C.H.P. unit   = 0.5p/k.Whs

   d. Normally quoted
      Cost of Gas   =   33p. per Therm
      It is significant to
      transpose the cost of Gas
      to p/kWh. Since there are
      29.3kW/Therm the Cost of
      Gas can be shown as   1.13p/kWh

From the above it can be seen that the claim that C.H.P. will cut fuel bills in half is a valid one.

13. Figure 3 shows a typical spread of unit capital costs as a function of size e.g. a 90 k.W unit will cost about £45,000.
   Figure 4 shows various manufacturers' maintenance costs as a function of electrical output.

Feasibility Studies, based mainly on fuel costs over the last year, can be produced quickly using a computer programme. This will give the size, number, cost of the system and pay-back period.
Pay-back periods of 2-3 yesrs seem to be currently acceptable. But if this capital outlay seems too expensive the alternatives are leasing or, if suitable and available, converting stand-by generators.

14. Recently the Government has cut its spending on promoting energy efficiency. In 1986 the Energy Efficiency Office (E.E.O.) received £24.5 million. This year, 1990, it will receive £15 million. The good news is that the European Commission should have agreed funding under a new energy-efficiency promotion programme known as Save. This will mean that there should be a lot of money available in U.K. for energy efficient schemes like C.H.P.

15. SOURCES OF INFORMATION

Books on C.H.P. are difficult to find in shops and Libraries. Even such an excellent source as the Institute of Mechanical Engineers has little or none. Two available are by R.M.E. Diamant:-

   a. 'Energy Conservation Equipment'.
      Published by The Architectural Press.

   b. 'Space and District Heating'.
      Published by Iliffe Books which has an informative part by J. McGarry on the Fundamentals of Heating Engineering.

16. The best source of literature, specifically on C.H.P.,can be obtained from the Energy Efficiency Office (EEO) which has published Good Practice Guide No.1 : Guidance notes for the Implementation of Small Scale Packaged Combined Heat and Power.
Also available from the EEO are:-

   a. Energy Manager - a periodical which gives the addresses of Regional E.E.O.'s.

   b. Case Studies of C.H.P.

   c. Expanded Project Profiles on Energy Conservation.

Information is also available from:-

   a. British Gas which has just published 'C.H.P. - The Way to Lower Your Energy Costs'.

   b. Energy Technology Support Unit (ETSU)

   c. Combined Heat and Power Association (CHPA)

   d. Association of Independent Electricity Producers (AIEP)

17. TERMINOLOGY

a. HEAT/WORK DONE/ENERGY

1 B.Th.U is the amount of heat required to raise 1 lb of water through 1° F.

    1 Therm = 1000 B.Th.U.
    1 Therm = 29.3 k.Wh.
    1 Nm = 1 Joule

b. POWER is the rate of doing work

    1 Nm/sec = 1 Watt = Standard Unit of Power

    1 HP = 550ft.lbs/sec = 746W = .746kW

c. PRESSURE

    1 Atmosphere = 14.7 psi = 1.01 bar
    1 bar = 10^5 Nm/m^2
    1 psi = 28" of water gauge
    1 m bar = .4" of water gauge

d. TEMPERATURE

$$T^\circ F = T^\circ C \times {}^9/_5 + 32$$

e. CALORIFIC VALUE of a fuel is the energy released during the complete combustion of unit quantity of the fuel. A fuel containing hydrogen can be said to have two calorific values. Higher or Gross C.V. and Lower or Net C.V. which is obtained from the Gross C.V. by subtracting 2442 kJ for each Kg of vapour formed

              OR

L.C.V. = H.C.V. - heat obtained by cooling or condensing water vapour.

f. PREFIXES

    k - kilo = 10^3 (a thousand)
    M - Mega = 10^6 (a million)
    G - Giga = 10^9 (a billion)
    m - milli = 10^3 (one thousandth)
    μ - micro = 10^6 (one millionth)

CONCLUSIONS.

Given the right circumstances, C.H.P. is definitely a money earner with one of the shortest pay-back periods in Energy Conservation Equipment. The case for C.H.P. is ideally summarised by a quote from a paper given by Roger Briggs at a seminar sponsored by the Institute of Hospital Engineers in December 1989:-

a. C.H.P. has tremendous potential

b. Don't be too enthusiastic

c. Make sure it's viable

d. Do it

e. Do it right!

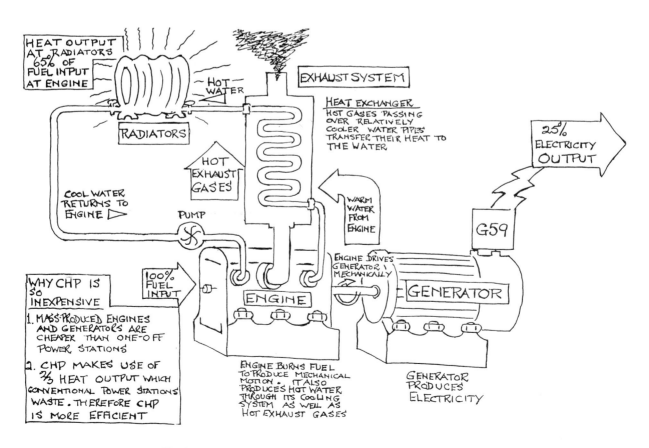

Fig 1   Inexpensive electricity from combined heat and power

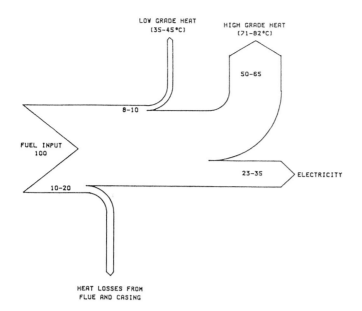

Fig 2   Energy balance of typical CHP unit

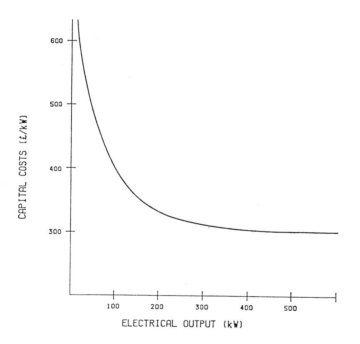

Fig 3   Installed capital cost as a function of size for gas
         fired spark ignition units

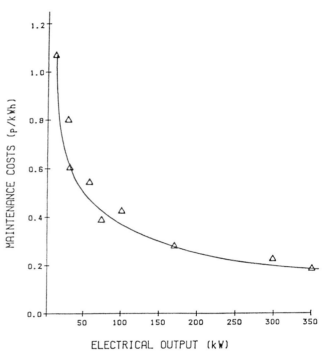

Fig 4   Average maintenance costs

The first 10 years on the UK market for packaged co-generation plant—a review

C J LINNELL
Applied Energy Systems Limited, UK

SYNOPSIS
This paper discusses the lessons learnt whilst also looking at the future for on-site Combined Heat and In-house Power Systems (CHIPS).

The great missed opportunity of the 80's!

Environmentally friendly

Free electricity or heat is finally acknowledged in 1990.

A review of the first ten years of packaged skid mounted mini- power stations has concluded that the 1980's saw one of the greatest missed opportunities. Proven, cost effective and environmentally friendly mini power stations here-on-out, referred to as combined heat and in-house power systems "CHIPS", have been over-shadowed. The unobtrusive features of CHIPS, coupled with the sheer simplicity and legality of them have given the technology an extremely low profile. The power potential of wind, wave, tidal and other renewable power sources, (generally more expensive, less available and less reliable) have rather stolen the lime-light. Also this has been helped by the electricity supply industry, who knew for certain that they posed little real threat to the status quo.

However, the proven and immediately available cost effective solution offered by CHIPS is a <u>very</u> real threat, especially at such a sensitive time as now when we have privatisation of the electricity industry. The true cost of nuclear power has only recently been advised and the subsequent acknowledgement of CHIPS has only come about at the beginning of this decade. **CHIPS could, theoretically, produce all the electricity requirements of the UK.** A very powerful statement one might think but the prospect of a diversified and integrated power production system/network would appear closer today than at any time in the past. The increase in electrical generating capacity or the replacement of time expired plant can certainly be covered by CHIPS. To continue to build power stations, even of the combined cycle type, even run on gas, cannot be justified, and is not only an offence but a direct challenge to the environment as well as to the moral and economic welfare of the world.

Environmental bonuses of better than 70% CO_2 reduction in emissions and additional benefits of energy savings to help conserve world resources are available. Further more **CHIPS provide the lowest cost heat and electricity to the end user.** This is because it is produced on-site, or adjacent to the heat need, thus providing the customer with consequential savings in distribution losses, and the avoidance of any inconvenience, or cost that the massive infrastructure would otherwise have incurred.

The UK Government has just released reports upgrading the previous predictions for CHIPS from 1-2000 MW's increased capacity for the 90's to 30-50,000 MW's by 2020, some change!
As more and more units are installed, the economics of CHIPS are continuously improving so much so that the installed cost per kWe decreases considerably. This in turn enables a significant increase in market penetration enabling units to be installed on sites where there is less utilisation but still nevertheless subsequently providing attractive paybacks, even taking account of the reduced hours of operation.

With the prospects of a carbon tax and even environmental credits looming, coupled with a social conscience and EEC legislation, the enthusiasm for the technology has snowballed and the outlook has never been so good!

British gas tariffs have been structured in such a way as to be beneficial to CHIPS and have now acknowledged the benefits of securing and stabilising gas loads.This can be seen by extensive CHP advertising for the first time which started in December 1989.

Perhaps the most significant change of all is the change of position adopted by the electricity distribution companies. Their established business above 1 MW is now under direct competition from the descendants of the CEGB and other new generating companies, and in 4 years time the business of customers of 100 kWs will be up for grabs! The solution of installing and operating CHIPS on the customers premises, and selling the electricity and heat, is being seen as a major opportunity to retain load and to operate in an unregulated but related market.

The development of new multi-fuelled engines for the truck and transportation markets to meet proposed emission legislation will provide the next generation of engines for CHIPS, leading to prospects for increased electrical efficiency and possibilities of direct injection of liquid gases and other new fuels.

The utilisation of the heat from the packages will subsequently be developed to produce not only chilled water for air conditioning but also temperatures down to minus 40 centigrade. Systems have already been demonstrated but commercialisation and cost effective production has yet to be achieved. The use of desiccants for dehumidification and also for cooling, using the heat from the CHIPS for regeneration, is in a similar commercial situation.

At the beginning of the 1990's CHIPS have finally been acknowledged and world market prospects are exceedingly encouraging.

HISTORICALLY

The production or capture of useful power from wind or water has allowed mankind to develop. The industrial revolution was said to have been driven by steam, with internal combustion engines coming into use fairly early on. Far from being a new technology, CHP is in fact a re-birth of sound engineering practices from days gone by when there was no alternative. All early industrialists had to be independent power producers on their own sites. However, as machinery developed and larger systems became more efficient and cost effective through buying fuel cost effectively, the benefits of large scale power production brought into being the large thermal power stations, which finally were connected by the national grid. This has given the UK one of the finest power production and distribution systems in the world. The best thermal efficiency of the large central power stations is in the order of 38-40% lhv at the power station. With the new combined cycle plant offering potentially 40-50% lhv, distribution losses can be high and are rarely less than 10%. This results in an average delivered efficiency of 26-28% hhv with the remaining energy being wasted to the environment.

It was against this background that CHIP systems were developed, providing delivered electrical efficiencies of 23% for small units (less than 30 kWe) and up to the low 40% hhv for compression ignition engined units (around 1 MWe). Overall energy efficiencies for CHIP systems

range from the mid 70's to better than 95% hhv.

The cycle of events has turned again with the lowest cost heat and electricity now coming from on-site heat led generation. This has become possible with the advent of low cost electronic controls and protection, coupled to reliable and efficient prime movers. It is possible for even the smallest units to be operated totally automatically at very high overall thermal efficiencies. The cost of delivered electricity produced by CHIPS ranges from zero to approximately 2p per kWe hr. This depends on fuel costs and the value of the heat, this assumes the unit is heat led and is not used if there is no requirement for the heat. This does not preclude the use of the plant with a heat dump facility. However, the case for the plant maybe but not always weakened as there are many circumstances when this operational philosophy still remains sound. CHIP units also have a lower capital cost of £350-600 per kWe installed than large central stations which range from £600-1200 for a coal fired station. It must also be remembered that we need to take account of the costs in relation to the distribution infrastructure and distribution losses. Additionally, delivery time for CHIPS is a few weeks as opposed to some 10 years for a large central coal fueled power station. This enables the installed generating capacity to be increased incrementally with demand and does not require long-term forecasting which has proved so difficult in the past.

MARKET DEVELOPMENT

The electricity supply industry have, to date, successfully suppressed the prospects for packaged CHIPS. The CEGB programme on alternative electricity generation, specifically excluded packaged CHP from evaluation. The reason? The only answer to this question, controversial though it may be, is that packaged CHIPS are the only proven alternative, providing potentially substantial competition to the electricity supply industry (ESI). However, the cost benefits to the end user are extremely attractive. The only significant support for this technology has been from the Department of Energy, (Energy Technical Support Unit, ETSU). Their budget for the promotion and demonstration of CHP has amounted to a mere few million pounds. When one compares this to the expenditure on the alternatives of wind, tidal and wave power, it is a quite insignificant amount. Also, when compared to the amount of money spent within the nuclear industry, this figure becomes even more insignificant. The fact is that a few million pounds spent on CHP P.R. and Marketing may mean that the technology has now been accepted by some friends and supporters. However, this is not enough and the right people have yet to be successfully lobbied.

The newly privatised British Gas Company, having unravelled the monopolies and mergers commission requirements and those of OFF Gas, in December 1989 (the end of the first decade of packaged CHP units) took full page advertisements in the quality national daily and weekend newspapers. The sales and marketing teams of British Gas are now gearing up to meet the challenge of the privatised electricity supply industry.

Historic pricing polices of British Gas had previously restricted the use of spark ignition gas engined CHP systems to the smaller installations that used firm gas supplies, as these units could not burn heavy fuel oil. The price of the gas supplied was related to the alternative fuel that the plant could burn. The changing of this policy, in May 1989 and December 89, has subsequently allowed for larger schemes to progress forward.

It is our view that future electricity generating capacity should be from mixed and diverse sources, and the 1990's will see the outcome of privatised ESI and the true cost of the electricity it supplies.

CHIPS always deliver the lowest cost heat and electricity to the user. The ability to turn **yesterdays waste into todays energy** has many attractions, one of which is that it improves overall efficiency and uses less resources.

To ignore this cost effective, environmentally friendly and energy efficient system will be to the detriment of the customer /end user. The environmentalists and opposition politicians need to educate the consumers.

More than 400 units have been installed in the UK in the last decade with most of the units using pipe-line natural gas as fuel, with sewage gas being the next most frequent alternative. The unit sizes range from 15-1900 kWe, all skid mounted, with a 6 MW combined cycle plant being the largest reciprocating engined installation during this period. The majority of units have been induction generators principally operating in parallel with the grid, most commonly connected at 415 volts. Operating automatically the units respond usually to the need of the heating system for heat. If the gas were free then the operation would be controlled by either the need for or when the best value for the electricity could be obtained. The principle use of the system is to simultaneously produce heat and electricity for the user at a cost advantage. Most of the units installed over the last 10 years were of the spark ignition gas engines kind, most being of the stationary industrial class. However, in the early days, converted automotive engines, many from Fiat, were used in the 15 kWe Totem package. The payback period for these packages was normally in the 2-4 years range. Most of the early systems were installed as base load units, which generally provide the shortest payback. More enlightened

customers are now installing units on the basis of achieving the lowest long-term running costs, which can mean that the units are not fully utilised and may be used for as little as only 3000 hours per annum or less. The marginal cost increase necessary to install a CHP system instead of other equipment, such as a new or replacement boiler or stand-by generators, often achieves a short and attractive payback without having to do many running hours. The last decade has seen the beginnings of many changes but most important **a change of views.**

In the beginning CHP systems were looked at only as an option - if there was time. Now customers are asking for CHP systems to be evaluated. The more advanced building services engineers and energy conscious consultants are looking at projects to see if there is any reason not to install CHP; a complete reversal of the previous situation! It has become quite a difficult task to provide sound reasoning against installing a CHP system. In order for the costs to be fully assessed, leading to the selection of any form of energy generation, whether it be heat, cooling, or electricity, both the load profiles and loads need to be fairly accurately predicted. However, this has been shown to be a fundamental weakness due to the lack of available data. This has come as quite a surprise to many, although it should not have; after all, the previous requirements of most customers was to have sufficient heating and cooling together with a sufficient electrical supply. The customer paid for the usually oversized plant; the consultants, contractors and plant suppliers all sold more equipment and the customer picked up the bill and continued to pay for the lifetime of the plant, usually operating less efficiently than more closely sized plant would have done. The change in emphasis to having sufficient but not oversized plant with improved controls via building energy management systems, means that the actual energy use and profile becomes a fundamental point. The electricity tariff, service capacity, maximum demand requirement and the emergency generating capacity all need to play a part as to the sizing and payback of the CHP system.

If one were prepared to wait for the often reluctant Area Board to respond the connecting of private generators was always possible both technically and commercially. In 1975 the ESI produced a guidance document, G26, thereby providing the technical framework for the connecting on of private generators to the system at voltages above 600 volts. In 1981 guidance document G47 was released, setting out technical requirements for the connection of private generators operating at low voltage in parallel with the area board. These documents were later combined and up-dated into G59.

The 1983 Energy Act saw for the first time an obligation on the part of ESI to, adopt and support CHP and publish tariffs, showing no favour or disadvantage to the private generator. The tariffs were published in October 1983 marking a major turning point in actually commercially marketing CHIPS. Prior to this the time taken to process a scheme made it practically impossible to commercially sell such systems. The problems of marketing the alternative home produced electricity from combined heat and in-house power systems then became apparent.

The Customers

Having been told for so long that it was neither possible or legal, and a complicated affair, to operate CHIPS, customers found it very difficult to believe that all had suddenly changed. A catch 22 situation arose, as if a change had taken place causing a complete about-turn of previous thinking could this not happen again in reverse? However, in the event the ESI failed in their obligations to adopt and support CHP in a number of ways.

Customers were not informed of their rights nor of the potential saving opportunities. It was left to customers to make the initial enquiry, who then rather took pot luck as to whether they spoke to a knowledgeable CHP person within the Area Board, whether it was known CHP was even allowed or whether details relating to commercial conditions were generally known, except, of course, at much higher levels. The lack of education of Area Board personnel led to a great deal of confusion and misleading stories being circulated. Some of the problems included penalty standby charges and extremely expensive export metering. It has been amazing that a completely negative attitude by the ESI has been allowed to continue. Perhaps it should not have been such a surprise if one considers the fact that private generation is, of course, against their commercial interest.

Compared to other countries with CHP, or co-generation (as it is known outside the UK), the UK tariffs accurately reflect the influential power and position of the established ESI and do not reflect the realities and merit of CHIPS. The simple laws of physics seem to have been conveniently forgotten! For instance, whoever heard of sending a package of electricity from one place to another several miles away, when there is more of a load in the immediate area where in reality it will actually be used. Electricity flows very much like water in many respects and perhaps would be better understood if considered that way. The present tariffs always assume that the electricity of private generators is of a negative impact on the distribution system, whereas the reverse is in fact the norm. Together with an enormous loss of opportunity for CHIP users and with the associated case for energy conservation, lateral thinking, in this particular case, lost out to the historic thoughts and protection of the status quo.

Tariffs for CHIP Sites

The concept that there should be any difference in tariffs for operators of CHIPS is the wrong stance to take, as no other electricity customer has any difference made for them.

If a customer selects an MD tariff, uses very little electricity or has occasional usage, or reduces consumption by conservation or by process improvements or by closing down part of the facilities, there are no specific penalty clauses. Bearing this in mind, the question must be asked why, therefore, should there be any different tariffs for CHP operators.

The tariff allowing private generators to export electricity from their own facilities via the grid has, for the past seven years, been both complicated and heavily weighted against the private generator. The coming of privatisation of the ESI should provide a more level playing field. The main generating companies will have to use the same formula for pricing and losses as the small private generator. This is always assuming that the laws of physics are not completely disregarded this time !!. The regulator at (OFFER) The Office of Electricity Regulation should be the arbiter that oversees fair play, which the Department of Energy has failed to do in the past. This is the only way that any real competition in the market for electricity can be achieved.

CHIPS size range and engines

CHIPS are available in skid-mounted units ranging from 18-1900 kWe and larger units will become available with installations up to 10-12MW electrical output becomes possible. The size of the packaged CHIPS is always referred to by its electrical output as the amount of heat recovered can vary depending on the temperature that can be used at any time or with any specific installation. The dumping of heat, although not encouraged, can frequently be cost effective. The commonly used term 'heat to power ratio' is subsequently a misleading term and should thus be avoided.

The stationary spark ignition gas engine is frequently chosen as the prime mover of CHP as it provides the lowest life cycle maintenance costs, the lowest engine down-time, the smoothest running and a low noise level. The alternatives are automotive conversions at the smaller sizes or compression ignition engines, ie. diesel, heavy oil or dual fuel.

The use of gas turbines is usually restricted to set sizes above 3.5MW although units are commercially available from 400 kW. However, the low shaft efficiency and the capital cost along with the rapid worsening of the shaft efficiency at part load, leads to these machines only being considered when the heat from the jacket of the reciprocating engine (up to 125 deg C) cannot be utilised and when the fuel is being bought. Otherwise the higher shaft and usually higher overall thermal efficiency, wins for the reciprocating engine.

The higher thermal efficiency arises from the ability to recover the latent heat of exhaust gases. The dew point temperature lowers as the excess air rises in the exhaust gases, with gas turbine having the highest levels. The lower the dew point, then the lower the useful temperature of the heat. The other important point is that the higher the excess air flow, the more heat leaves the system without the energy being recovered if there is no condensing. Gas turbines have the majority of heat in the exhaust. The performance of a the gas turbine is adversely affected by increasing the exhaust back pressure by using heat recovery heat exchangers, these have to be kept to a minimum pressure drop which tends to make them larger and more expensive. There have been some occasions when gas turbines have been used contra to the above reasons. Reciprocating engines are some-times viewed as toys and gas turbines are for real engineers !!

If properly maintained, the design plant life of CHIPS can be over 20 years. The increasing interest in exhaust emissions has further strengthened the reasons for gas fuelled systems (methane) as they offer the lowest level of emissions of any fuel other than hydrogen. The relative price stability offered by British Gas is a further incentive in favour of gas.

Electrical efficiency varies from the 18kWe machine at 23% higher heat value 'HHV' to the low 40's for the slow speed large compression, ignition engines. The exhaust gas temperatures vary in the range of 400-600 deg.C. and it is therefore possible to produce steam at pressures above 100psi. However, this is usually restricted to larger installations for not only practical but also for cost considerations. Engines of 300kWe and above can produce steam from their jackets, although the pressure is restricted to 15 p.s.i.g. Otherwise the jacket water is between 95-112 deg.C. and the heat from the alternator and the engine casing and associated pipe work is available at 35-40 deg.C. If this heat is not used, then it has to be removed. The highest thermal efficiencies are achieved by using the heat down to the lowest possible temperatures.

Examples of the application of the technology

The first of over 285 CHIPS were installed in swimming pools at Ormskirk and Skelmesdale in May 1983, and provide 23.8% electrical efficiency with 40kWe induction generator CHIPS. These units operate at an overall thermal efficiency of better than 90% with the high grade heat used to displace the heat otherwise produced by the boilers. The low grade heat from the condenser is used for pool water heating. The electricity is fed onto the main bus via the normal fused isolator which can be locked in the 'off' position only. The gas engine uses the induction motor to start it and then, gradually, as the horsepower is increased, usually in about 1 second, the motor is turned and operates as the generator. The commissioning of both these units was witnessed by the Area Board (and electrical protection settings demonstrated.) These installations were successfully IMPLEMENTED by turnkey contract by Applied Energy Systems Ltd. The site personnel carried out the normal daily maintenance with the local Home-Tune Service man undertaking monthly services. The physical distance of over 440 miles round trip precluded the units from being economically serviced by AES. However, as the number of units increased nationwide, full contract maintenance has become both available and cost effective to both parties.

Another AES turnkey installation at Redhill Hospital, 1 off 90kWe CHIP system, was commissioned in 1986, which achieves nominally better than 92% fuel efficiency whilst generating electricity at 25% efficiency hhv. This has been achieved by the full utilisation of the jacket, oil and high temperature exhaust heat. In addition, the exhaust gases are further cooled by the use of a condensing exhaust heat exchanger which brings the exhaust gas leaving temperature down to 10-15 deg C. This additional circuit is pre-heating the domestic hot water. It can be seen that this type of application can be used in most large residential applications, including hotels, group heating schemes and anywhere significant quantities of water have to be heated from cold.

The installation at Copthall Leisure Centre consisting of 2 X 90kWe CHIPS, commissioned in 1988, achieves in excess of 95% overall thermal efficiency hhv. This system replaced 4X4 electric heat pumps which were switched off in 1986 being too expensive to operate, when compared with gas boilers. Increased operating efficiency has been achieved by the use of all the previously described techniques and the further use of ventilation air from the acoustic enclosures. The plant room air is nominally 33 deg.C. entering the enclosure, where the temperature is raised further by the engine, and the air after the heat has been recovered by the run around coil is 27 Deg.C. The run around heat recovery circuit puts the heat into the incoming pool hall ventilation air.

A similar system utilising a 40kWe CHIP can be seen at a dry sports centre, and a 90kWe CHIP has been installed at Howarth Scouring Wool Mill. This particular unit is used principally to heat washing water with the hot air being used for a wool dryer.

During a 1986 refurbishment at Reading Central Pool 3 x 40kWe CHIPS where installed which also included export meters. This has enabled the exported electricity costs to pay for the standing charges and any extra electricity that has been used, with the on-going effect of the customer not receiving an electricity bill for some two years. Gas consumption was also reduced and as stated the export electricity contributed towards paying the gas bill. This installation also won a British Gas GEM Award in 1988. This local authority now operate a total of 9 X CHIP systems, 4 being installed into a new build leisure centre and 2 into the Meadway Sports Centre during refurbishment work, also utilising export meters.

Chichester was the first of the new build leisure centres to incorporate CHIPS from original design. The 2 x 40kWe CHIPS were originally specified as 3 units and the third unit was installed three years later following the successful operation of the first 2 machines. The same thing subsequently happened at the Kingstanding Leisure Centre. Both of these installations won area British Gas Energy management awards (GEM) and altogether CHIPS have now featured in 18 such awards.

Twenty-five new-build leisure centres now have CHIP systems as original equipment and such systems continue to be specified for new build leisure centres. The practice of increasing the number of CHIPS on site after a successful initial period, has also been adopted in many other places, one of which is Biggarts Hospital in Scotland. At this site 2 x 40kWe induction generator CHIPS were increased to three. Export meters were installed as part of the original specification recording the exported kWs. This installation won the 1988 Scottish Gas GEM award. Further information can be found in ETSU Case Study No.17.

The largest installation to date in any leisure centre has been 3 x 90 kWe CHIPS at the Magnum Leisure Centre in Irvine Scotland. This installation gives a total of 270kW of electricity and 630kW of heat whilst operating at 92% plus thermal efficiency, at the risk of sounding boring, this installation produced another award - the 1989 Scottish Gas GEM Award and the first British Gas national small scale CHP winner in 1990.

An ETSU Case Study (number 5) is available which outlines the history attached to the 2 x 90kWe CHIPS installed at the Swiss Cottage Leisure Centre. This makes for some interesting reading as the units, operating at 92% overall thermal efficiency, are in a unique installation, with the units using a soft start system. The intention originally was to reduce the starting current of the large induction motors, frame size 105kW, this being considerably larger than any other motor on this site by a factor of 3. The outcome of using the soft start was to provide a slow steady start. However, this did not reduce the absolute amps drawn, but only prolonged the period it was drawn for. Subsequently Star Delta starters are now used for units up to 100KW and larger units are generally starter motor assisted. Reduced current

start is achieved by the use of power factor correction capacitors which are either bought in in stages or in one unit dependent on size, before the star contactors are closed. This in turn reduces the in-rush current by the rated capacity of the capacitors by providing some of the reactive current element.

Group heating schemes

Whereby wet heating systems are installed afford an excellent opportunity for the addition of CHIPS. There are a number of schemes in operation and one of the first was at the Bakers Alms Houses, within the London Borough of Waltham Forest. Both social and economic benefits are demonstrated by this installation, as again an excellent Case Study by ETSU (No.21) will prove. The scheme is based around some 52 two-up-two-down terraced houses configured in a 'U' shape. The result and application of CHIPS has been that tenants no longer pay standing charges for their electricity supply. The unit rate has come down by nearly one full pence per kW hour. Pre-payment meters have been installed rather than the customer operating an account system and heating bills have been reduced by #1.70 per week. This is a direct result of the installation of the combined heat and power system, part of a housing improvement scheme at present in operation by the Council.

The electricity for the scheme is produced principally by CHIPS and any exported electricity is sold to the LEB as the local Area Board. The additional top-up units are imported from the board on ''the use of systems tariffs''. This scheme is one of the first and smallest users of this tariff.

A further project is at Bulmer Court consists of 56 dwellings established with three blocks of flats. These installations are the forerunners of many more for the London Borough of Waltham Forest. Another interesting point is that these machines are operated by Waltham Forest Energy Services Ltd, a separate operating company set up by the Local Authority. The use of these systems in tower blocks, as well as additions to existing district heating schemes, can provide for a considerable reduction in operating costs and/or improvement in the level of heating affordable by the tenants. As can be seen, this improvement is a very important factor the alternative has been to use traditional electric heating. Many users with poorly insulated property have found electric heating considerably expensive and quite frankly beyond their means, hence their level of heating has been much lower than required. Consequential damage to property has been increased by condensation which in turn causes premature degradation both decoratively and structurally, the cost of which has yet to be evaluated.

CHIPS as standby generators

A CHP installation of the 140kW frame size can be seen at the Devon and Cornwall Police Training Centre at Middlemoor. Here again, ETSU have issued a detailed Case Study (No.15) on what has been the first unit installed into any police facility. The CHIP provides a large proportion of the heat and electricity for the most part of the year. An additional benefit is that the unit operates as a standby generator in the event of the mains power failure. If the machine is on-line at the time of mains failure, the unit will attempt to stay up and running by disconnecting itself from the grid in nominally 110-130 milli-seconds. If this is successful, then whilst there will have been both voltage and frequency disturbance, this sometimes allows the pump and fan contactors to stay in and small computers and controls still to operate. Therefore this provides for a crude uninterrupted power supply. If the unit is off at the time of failure, the unit will come on line as a normal standby generator. Several police stations and hospitals are now being equipped with CHIPS to operate in this way with units from 26kWe being selected.

Generating Electricity During Peak Value Hours

The use of CHIPS to generate electricity during the most advantageous times, for either the displacement of incoming electricity or for the best price for export electricity, requires that controls be used to make heat loads available. This can be achieved by the use of specified special heat stores or more normally by the use of the existing hot water storage plant or the thermal capacity of the building. By modifying the existing heat load profiles in this way increased installed generating capacity of CHIPS is achieved which therefore gives way to lower long-term running costs.

The sewage industry has, through the use of biogas produced from the digestion process of sludge treatment, been generating electricity for more than fifty years. The introduction of automatic unmanned skid-mounted CHIPS has provided increased opportunity for cost effective generation, and the use of existing gas storage plant associated with the installation thus allows the generating set to operate 17 hours per day with increased output at peak times if and when required.

The use of landfill gas, a by-product of the landfill acting as a digester, is a mixture of methane and many other gases, all of which continuously vary in their concentrations in response to either the age of the tip and the different material in the fill and also the weather. This is the worst fuel for engines. At least, the production of electricity uses the methane usefully, which would otherwise have acted as a potent greenhouse gas, and reduces the atmospheric pollution at the same time as helping to remove the dangers of explosive gas from the tip and to partially remove the associated odour. Most of the electricity produced at such installations is exported to the grid as many sites have little associated electrical demand.

The use of coal mine drainage gas, to fuel spark ignition gas engined power generation plant, which have a varying methane quality, is, at the present time, seeing a revival in response to a reduction in production costs, brought about by displacing bought-in electricity. Also this has the additional benefit of reducing the emission of methane to the atmosphere. The production of coal releases methane and as it becomes explosive at 4% this has to be removed for safety.

Air fuel ratio controls

The use of varying calorific value gas in engines has led to the need for an air fuel ratio controller which are able to accommodate large fluctuations. This requirement, coupled with the need for emission control and performance correction between services has resulted in AES developing a cost-effective unit which is a multi-function multi-mode air fuel ratio controller. This has proved to be extremely reliable and is shortly due to be marketed.

CHIPS can produce chilled water as well as heat

CHIPS can also make use of the heat off the package to drive an absorption chiller to produce chilled water for air conditioning, or to process chilling with temperatures as low as minus 40 deg C being attainable. This technology produces many benefits not least of which is that it is a cost effective way of producing chill whilst not using electricity. This also produces a dramatic reduction in the green-house effect by the subsequent reduction of the polluting gases which also shows up a high level overall energy efficiency. The Hotel De France in Jersey is using 4 X 100kWe diesel fuelled CHIPS and a 60 ton Carrier absorption chiller as the base load plant for the provision of the building services energy.

Combined cycle plant at North West Water Ltd - Davyhulme S.T.W.

The largest installation of skid-mounted CHIPS can be viewed at the Davyhulme STW, part of the North West Water Ltd where the application consists of a 6MW combined cycle plant. Three 1900kW Waukesha Dresser lean burn spark ignition gas engines provide the waste heat to produce 16.5 bar absolute steam at a total of 5 tons per hour. This is converted into electricity in a 310kW Dresser steam turbine. The pass out steam at 1.3 bar absolute and also the heat from the jacket, is used either for the digester heating or dumped. The main induction generators are 6.6kV operating into the

NORWEB grid. The sewage gas fuel previously was used in up to six 1200kW dual fuel engines and the new installation now enables the production of 3MW of exported power after satisfying the site demand. The power house refurbishment was carried out as a turnkey contract by AES Ltd, the old plant being kept online until such time when the new plant was able to take over virtually without any shutdown. This was a very significant point, as Davyhulme STW handles a large amount of the sewage from Manchester. The new plant will operate, in the main, unattended with a fully computerised control and data acquisition system, thereby not only releasing manpower for other duties, but also improving the cost effectiveness of the plant.

North West Water Ltd, in updating a number of plants elsewhere, have a couple of sites which have been green field sites whereby the power house has had to be built. Oldham S.T.W. has two 315kW Waukesha VGF24GL lean burn engines with a low pressure gas mixture being fed through the turbo charger. Induction generators are used on both units, as well as at Dukinfield which has three 315kW generator units installed under turnkey contracts, subsequently proving to be increasingly advantageous. The use of induction generators is normally kept to below 100KW. However, the simplicity and the ruggedness of this type of machinery and controls has since found an increasing number of applications.

Summary

This brief review of CHIPS demonstrating power generation in parallel with the grid should leave one in no doubt that the packages **do** exist and **are** being adopted more and more. AES Ltd as the market leader, have manufactured 285 CHIPS, 47 of which have since been exported to Japan. The Department of Energy (ETSU) have a significant collection of monitoring reports and independently gathered data which can be made available free of charge see the list of some of the documents. Other organisations, such as British Gas and the Combined Heat and Power Association are useful additional sources of data and the overwhelming evidence will, in the near future, ensure a massive surge of further installations both at home and overseas.

Reviewing other CHP skid mounted packages and packagers.

MARKET BACKGROUND

The first packaged combined heat and power units into the UK market in 1979 were FIAT TOTEM (although this unit was manufactured by BIKLAM of Italy and with Fiat having a minority share-holding the unit was perceived to be supported by the Fiat organisation as Fiat energy was the name of the first UK distributor). 110 of these 15 kWe Totems, including the unique self-excited induction generator types,

which could operate as standby generators, were sold by October of 1985 when they effectively with-drew from the UK market. By this time the distributorship had transferred to Associated Heat Services (AHS) Co-generation Department. The closing down of this operation came about when a new managing director was appointed to the AHS Board and the true financial position of the extensive loss-making department was made clear. In December of 1985 Applied Energy Systems Ltd was contacted by Biklam with a view to a possible take over by AES of the support of the existing Totems and future sales of the product in the UK. In April 1986, AES employed three of the AHS engineers, including their Service Manager and took over a section of the parts holding while at the same time helping to negotiate the return of 25 Totems to Biklam as these were surplus to requirements and technically unacceptable to AES. AHS subsequently closed the cogeneration department and the anticipated signing of the distributorship agreement with Biklam and AES never was concluded, as BIKLAM could not accept the terms and conditions and technical changes necessary for the product to continue in the UK market. In October 1986, a service company was appointed but by this time the full cost of operating TOTEMS was becoming known to the users and specifiers and the number of TOTEMS in service continued to fall.

AES provided the full technical support and training of the Bass transport service engineers who took over the service of the 50 machines within the Crest Hotel chain. The supply of spares then became a direct deal with Biklam and AES sold its remaining spares to Bass. The number of Totems in use today in the UK is unknown but some have run very satisfactorily, particularly those in sewage works where the maintenance was carried out by interested on-site staff; Obviously the direct maintenance costs are much less under such circumstances. These sites also served the purpose generally up-grading the installations and controls to the required levels and operated at return water temperatures of 65-70 deg c. The main failings of Totems were poor installation and lack of boiler controls which meant that the units did not run for the required number of hours. The pressure for low price installation and low maintenance costs caused all the importers of Totem in effect to subsidise the units. In Holland where over 650 of these units are installed the maintenance cost is quoted at about 50p/hr run, whereas in the UK, the final maintenance cost was 14-17p per hour run. However, the higher price necessary in order to perform the service correctly could not be supported in the UK by the price differentials between gas and electricity, whereas in Holland, this was not a problem. An additional problem with the performance of the units was the rapid fouling of the plate heat exchanger by the boiler water which is generally untreated, unregulated, make-up water.

Another factor was the failure of the

generator bearings on the units which had their cooling changed from the building services / boiler water to the engine jacket circuit. This was done to eliminate the problem whereby the small waterways in the generator cooling circuit were fouled by the boiler water, however, the increased temperature of the jacket water caused the premature failure of the bearings. The exhaust gas heat exchanger needing regular cleaning on the gas side proved to be a time consuming and therefore expensive operation. A new cylinder head was needed or was chosen as the most cost effective route for the requirement of a service overhaul at 5000 hours. Flexible water pipes needed to be replaced frequently due to the high temperatures in-side the box.

The control circuits in the TOTEM failed unpredictably which was put down to the temperature and the vibration to which they were subjected.

There were many other problems of a minor nature but problems costing money or time to put right, all detract from the profitable operation of any CHP package. The full service history of 108 TOTEMS installed in the UK up to July 1986 are held by Applied Energy Systems.

Serk Cogeneration units withdrew from the market as a result of BTR's appraisal that the products would not meet short-term financial requirements leaving 21 more units and customers without support.

18 kWe and 40 kWe Serk units were of both a basic simple and robust construction with only one main problem area being the electronics in the control panel. Even these were virtually eliminated at the time of shut-down of the department.

Again the pressure to achieve low prices to obtain very short paybacks caused the all too familiar problems of lack of boiler interface controls and the less than desirable installation short cuts. This resulted in some installations not achieving the rates of return expected.

AES secured the services of the General Manager of this department which served to continue to strengthen the AES team. AES has subsequently replaced 3 of the Serk units with CHIPS.

The liquidation of CLEAR Ltd in September of 1989 meant that 16 machines were left unsupported.

This company, from the outset was always looking to cut corners and to take the maximum return out of the package without regard for convention or the experience to support the ratings offered. This price-to-performance ratio was very highly geared so much so that there were no margins for wear/decline in performance between services, also the price

installed was low and the tendering procedures frequently used in the public sector make no allowance for employees lack of experience within the public authority. As a result this has meant there have been some unfortunate cases where the equipment has been left totally unsupported in the market-place, and where the companies who have genuinely given the better package have been pushed into second place where there is no prize, and only considerably more overheads/costs. The whole CHP market has been blighted by over-optimism, dreams, hopes, aspirations and predictions which have only resulted in disappointments.

A brief case study

Clear Ltd was the successful tenderer for the supply and installation of 3 CHP systems. 2 x 37 kWe and 1 x 55 kWe.

A question frequently asked is "WHY"?

If the CHP systems use engines, electric motors, heat exchangers, controls, cable, etc. in fact all readily available components, why then are they so difficult to produce, install and maintain?

Hopefully, the following notes will provide a brief insight into the answer to that particular question.

The conclusion must however be that experience and expertise are not always visible to the inexperienced and hindsight is an exact science.

These packages claimed more power output than they could actually deliver for this application long term. The design of the package was naive at best and incapable of achieving short-term reliability, let alone over the next 20 years.

In brief the principle problems were:-

* The continuous rating was exceeded
* The engine sump could not be removed without removing the engine;

* The exhaust connection to the heat exchanger had no flexible to relieve thermal expansion or vibration;

* The use of unsupported pipe causing leaks and failures;

* The use of unshielded or supported cable again causing failures;

* The use of inappropriate electrical connections causing the wires to fall off;

The lack of vibration

isolation for the control panel causing the components to work loose and also to fail;

* The use of low quality fitting and components totally unsuitable for the purpose intended and thus causing failure.

Machinery power output

The engine was a conversion of a Ford diesel engine into a gas engine by Power Torque Ltd of Coventry. The continuous rating of the 4 cylinder engine when fuelled on natural gas for power generation duty is 37kW shaft. The CHP unit was rated at 37kW electrical output.
The 6 cylinder engine is continuously rated at 55kW shaft. The CHP unit was rated at 55 KWe.
The claim for more power than the conversion company claim for their product, is obviously a potential major problem in that it may not be able to deliver the power continually and will therefore not have any spare capacity to allow for wear or the loss in performance between services.

Special limited conversions of engines should always be viewed with special attention, as long term support for your investment may not be available on a cost effective basis, or worse still not at all.

Items missing that are normally fitted for CHP applications:-

* Oil cooler;
* Crank case breather control system;
* Electrical driven engine water pump;

With the engine being sited on top of the oil tank it means that the oil tank cannot be removed as it is inside the base frame. This then stops the sump from being removed for the oil pump to be serviced, or the big end bearings, or the removal of the pistons, without the removal of the engine first.
On the 37 kWe machines the exhaust pipe is solidly connected to the exhaust heat exchanger and this does not allow for expansion or vibration isolation. This is not only likely to cause premature failure of the exhaust manifold, but also of the exhaust manifold gasket and the exhaust heat exchanger.On the 55kWe system the bellows can be seen without any thermal shielding.

The water cooled manifold on the 55kW set has already been replaced.

Access to the electric motor on the 37 kWe sets is restricted due to the position of the panel. The panel would have to be removed to replace the bearings or coupling which has to be done from time-to-time.

The control panel is virtually solidly mounted to the main skid which results in high vibration levels being experienced in the control panel and thus the premature failing of components

These therefore are some of the points to be aware of in order to avoid the traps just waiting for the inexperienced person to fall into.

This case study shows how three CHP systems, all meeting the tendering requirements, failed to deliver results.

Conclusion

The best protection for any purchaser is a proven track record of the potential supplier. Full references need to be thoroughly taken up covering the financial and technical position of the company, its intended suppliers, together with the best surety that can be gained. This generally means going with the established market leader. Even when first price is higher, there can be very good reasons for that. The fact that they make a profit will ensure your support being available. When the supplier a customer has chosen goes out of business, the lack of support for the package can mean a total loss to the organisation.
The purchase of a standard product brings with it the embodiment of experience, the machines have been produced before, and this is sound advice to follow in order to safeguard your purchasing decision. It is wise to think before writing a specification, no one would write a specification to buy a car or van which tells the manufacturers how to build the vehicles. This case study is not unique; it happens all too frequently; Lets hope that it does not happen again.

Other package suppliers

Watermota

A relatively small private British engineering company, specialist engine mariniser, which has produced less than 10 CHP packages since first entering in to CHP market in the early 1980's. The size range of units delivered are up to 50 kWe. The company has been through an extensive learning period and has fully supported its product. The lack of marketing and facilities have ensured that the CHP units have been a sideline operation and an interest area only for the owner.This product was the first to make use of a programmable logic controller and has demonstrated some of the highest electrical efficiencies of any unit of the 45 kWe frame size.The relatively high standards of the packages and installations all point to a potential supplier of a significant number of units in the future.

Holec

A UK company operating as the UK distributors for the Dutch company Nedalo BV.Size range of units delivered into the UK range from 18-300kWe. Since the early 1980's, a total of approximately

64 units have been supplied and the majority of units are being acoustically enclosed and of the self excited generator type, this has generally given them the disadvantage of being more expensive than the induction generator type units which had also been supplied without acoustic enclosures for most applications. As with all other companies Holec have experienced the pain of gaining experience together with the cost of supporting the widely distributed low density installations. At the beginning of 1989 the company introduced a range of induction generators along with the use of an on-board computer. The continued development of this product and the support services should put them in an extremely interesting position for the increased business that should become available.

Combined Power Systems

A UK company, born out of UMIST in Manchester, produce and delivered units in the size range up to 185 kWe based on the Ford BSD and MAN engine ranges. From the beginning the company was looking at the vast potential market and also the barriers. The company correctly identified many of deficiencies in the early pioneer packaged CHP units and the associated problems within this embryonic market.

Some of the problems were the poor reliability of the units, the lack of service contracts for maintenance, coupled with the lack of information as to the operational performance of the machines and also the lack of money to purchase CHP units. The additional factor relating to the perceived high price charged for the CHP packaged units has since proven to-be untrue, as CPS have now increased prices to market levels following an outstanding market penetration of 130 units in approximately 2 years of marketing on the back of low prices and some very attractive deals. However, the small print of the sales and service maintenance contracts provided by CPS can perhaps give one an insight into the thoughts of the company and product. Substantial assistance has been forthcoming from the DTI, BET, and a BES scheme, among others, has been the envy of the pioneers of packaged CHP units. The utilisation of the Department of Energy, Energy Technical Support Unit administered research and development grant scheme has been of major benefit and credit must be given to the Company for having made use of the scheme.

The remote monitoring data acquisition and control system is presently the state of the art as installed in any product. This radical and apparently cost effective system, used in conjunction with the maintenance expert system at the companies service department appears to have no equal.

The novel marketing approach of offering a Discounted Electricity Purchase Scheme, that is the service of converting the clients gas into electricity and heat for the payment on each unit of electricity produced, which gives CPS the income to pay for the unit and the maintenance and the capital cost of the unit has undoubtedly been of substantial assistance in the establishing of the companies product so quickly in the market place.

This company has publicly declared its ambitions of attempting to become the largest supplier of packaged CHP units in the world.

KFS Mini Power Station

A UK company best known for conversions of existing standby generators into CHP systems. The utilisation of the ETSU demonstration scheme gave significant publicity to the technology but to date there have been only a few replications. The idea of using existing plant that otherwise is not extensively used, is, at first thought a good idea. However, the potential pitfalls can be:
*   The engine may have been only suitable for low utilisation;
*   The unknown condition of the engine even though there may be few hours on the clock, will make calculation of future maintenance costs diffi cult;
*   The installation may not have been suitably posi tioned in order to allow for continuous running with regard to both noise and the difficulty of being able to connect the heat recovery into the heating system;
*   The exhaust may be suit able for use intermit tently, but not on a continuous basis;
*   The governor, along with the control and protec tion will need changing in most cases and the alternator may be unsuit able for use in parallel with grid;

If all this was not enough the fact that the system is, by its very nature, likely to be limited in number the future support of a non standard piece of kit together with the cost of producing maintenance information and drawings, all come together to take the shine off the original idea. At the end of the day the larger the machine, the more likely it is that a cost effective scheme can be put together. A total of 20 units making 4.75 mwe installed capacity. These include new gas and diesel units as well as conversions.

Dawson Keith

A significant generator set manufacturing UK company that has supplied a few CHP, units principally 3x300 kWe for North West Water Ltd for the Davyhulme STW and some 550 kWe units. Although the potential for further expansion into this market exists, however the commitment has not yet been seen.

Dale Electric

The largest English generator set manufacturer that, from time-to-time has produced and installed a few CHP units in the size range up to 300 Kwe. Notably these are the installation at British Gas, East Midlands of CHP units and an absorption chiller system, and also the installation of a couple of units operating at medium temperature with a heat store at the Cleveland Council Computer Centre. Additionally, as a packager of gas turbine power packages and manufacture of an extensive range of control panel systems they are well equipped. When Coupled with the existing extensive network of service support infrastructure, if the installation technology, support and commitment were more dedicated the potential for further involvement in the CHP unit market is substantial.

Shannon Power

A medium-sized generator set manufacture that has supplied CHP units up to 800 kWe. The intermittent presence in the market has resulted in 15 units in the UK and 7 units overseas the total number of installations. However, some of the installations have been at the leading edge of the application of the technology. Notably, the installation at the headquarters of British Gas North West which including a large absorption chiller and a medium temperature heat system is one such application.

Jenbacher

An Austrian engine company that has been represented by different companies during the last decade in the UK.The most notable of the few installations has been the 2 x 1500 kWe gas fuelled CHP units installed at Crossfields works (Coal Mines) in Liverpool in the early 80's. The two spark ignition engines are of additional interest in that it was a novel application of heat recovery which is being used for vacuum deaeration of the feed water for the boilers.

Brons MAN

Brons Industries manufacturing facilities was bought in mid 1989 by Dresser Waukesha Engines and the manufacture of the Brons MAN engine was ceased at that time. There are four engines in the UK.

Dorman Diesels

A UK engine manufacture (part of GEC) has updated its range of engines and has heavily invested in the latest technology particularly gas engines with a range of units that will shortly reach the 1MW at 1500rpm. The work on low nox engines at Dorman appears to place them well placed to take advantage of the developing CHP

packag market.

Leverton

This substantial company, part of Unilever Group, is also one of two Caterpilar engine UK distributors and has an extensive service and spares network. During 1989 the Company began to market CHP units based on these engines. If the Company makes available not only resources but also the commitment to this business, a significant market share could be secured. The worldwide reputation of this engine company and its success in Europe and the USA means it has to be taken seriously.

Many other companies have produced the occasional CHP unit. This review has only attempted, in the time and space allowed to cover those with any significant sales volume (over 20 units), or of some technical note, such that any interested party can further investigate the subject.

Summary Of EEO Publications

Project Profiles of CHP projects under Energy Efficiency Research & Development and Demonstration Schemes.

Case Studies of successful CHP Projects

Good Practice Guide Number 1 - Guidance

Notes for the Implementation of Small Scale Packaged CHP.

Good Practice Guide Number 3 - Introduction to Small Scale Combined Heat and Power. Available from ETSU (address below)

ETSU Study No. 8 - Potential for small scale CHP in UK Public, Commercial and Domestic Buildings. Available from Energy Publications, PO Box 147, Grosvenor House, High Street, New Market, CB8 9AL.

Energy Efficiency Series No. 5 - Combined Heat and Power and Electrical Generation in British Industry 1983-88.

Energy Efficiency Series No. 7 - Industrial CHP:

The Potential for New Users.

Energy Efficiency Series No. 12 - co-Generation and Electricity Production in British Industry 1988. Available from HMSO Publications Centre, PO Box 276, London, SW8 5DT.

Energy and Environment Paper No. 3.
Environmental and Economic Implications of Small Scale CHP
RD Evans
ETSU
for the Department of Energy

Energy Paper 58 - "An Evaluation of Energy Related Greenhouse Gas Emissions and Measures to Ameliorate Them." HMSO.

Recent Papers By C.J.Linnell Include Prospects For Small Scale CHP

Installations from less than 15kWe to greater than 10MWe. Presented at the 7th National CHPA Conference.

Package Cogeneration Systems - Less Than 3MWe - A Life Cycle Cost Analysis Comparison

Presented at The IMECH E Seminar Managing Engines For Profit.

Use of landfill gas in spark ignition engines

H D T MOSS, TD, MA, CEng, MIMechE, MIEE, MInstE
Shanks & McEwan (Southern) Limited, Milton Keynes, Buckinghamshire, UK

SYNOPSIS The first trials into the generation of electricity using landfill gas were completed using a gas carburetted version of the Rolls Royce 8 cylinder military engine. Subsequently, three 12 cylinder spark ignition gas engines were used to demonstrate the commercial viability of such a scheme. Problems with oil formulation and overheating are discussed and reference made to tariffs, noise and control. The success of the Demonstration Scheme should give confidence to other landfill site operators who control the environmental effects of landfill gas.

INTRODUCTION

Landfill gas is the product of the anaerobic decay of refuse which has been landfilled. Although it is regarded as being the product of the decay of municipal wastes, it can also be derived from commercial wastes such as paper and packaging. The United Kingdom produces some 13 million tonnes each year from the household or municipal sector but if the commercial wastes are included, the figure rises to some 25 million tonnes each year. Of this, the proportion which is landfilled is between 85% and 90%, the remainder being incinerated (by mass burn techniques), converted into Refuse Derived Fuel or into compost. The quality of landfill gas which can be derived from an "average" tonne of waste is variously quoted as between $200m^3$ and $450m^3$. Even if only $100m^3$ of this could be captured, it means that there is a vast quantity of energy available from refuse in the form of gas.

FIRST GENERATION

Shanks & McEwan (Southern) Limited operate a number of landfill sites in the Bedfordshire and Buckinghamshire Areas. Five of these are old brickmaking clay pits and were bought to the Group when Shanks & McEwan brought London Brick Landfill. Early work with landfill gas had lead to the first commercial use when in May 1981, it was used to fire bricks in a transverse arch Hofmann kiln at Stewartby Brickworks. The gas supplies were limited because of the landfill operation at the time so in 1983 work was planned to use the limited surplus of gas in a different way. Both West Germany and the United States had used landfill gas to generate electricity so a search was made of United Kingdom engine manufacturers to see if anyone was prepared to see this gas used on their equipment. Eventually the military division of Rolls Royce produced the answer with a gas converted variation of the B series engine. This 8 cylinder engine had run on test for a number of hours using natural gas and was thought to be capable of conversion to landfill gas.

Through a series of meetings, Messrs Fletcher, Sutcliffe Wild of Howbury undertook to link the B81G engine through a Newage gearbox to a Stamford 75kVA Generator. The gear box ratio was chosen to allow the engine to run at 2500 rpm and the alternator at 1500 rpm thus making sure that the engine could develop adequate torque for the purpose in hand. The units were monitored on a common bedplate and surrounded by an accoustic enclosure.

The unit was run on natural gas in April 1984, to prove that the equipment would function correctly. The only problem encountered was an oil seal in the gear box which was replaced before the unit was shipped to Stewartby in Bedfordshire. Because the unit was very much a prototype, there was a great deal of opposition to installing it as an extension to part of the London Brick 6.6kV and 415 volt networks. Consequently the control system was set up to allow the unit to run the landfill gas abstraction compound but not to generate into the local network. This meant the necessity for a changeover switch and a "black start" facility.

EXPERIENCE

The unit was commissioned in June 1984. The engine was started using propane and when it had run for long enough to warm the engine, usually 1 to 2 minutes, the gas supply system was switched from propane to landfill gas. The machine was run in the daytime only for the first six months. Oil usage was high, partly because of the oil usage in the engine. Eventually the gearbox fault was traced to an incorrectly machined bearing end cap. The engine oil loss was traced to the oil breather cap and although reduced by venting into the carburettor line, continued without further change. Eventually, a saddle sump was placed alongside the engine with a total capacity to allow 30 hours of unattended operation. Several 100 hour tests were completed after this without major incident.

Other problems occurred and although none was serious, the cost of maintenance and fault finding started to become excessive. Amongst the major items were frosting of the propane evaporator, sludging of the evaporator, magnetic pick up faults and unstable carburettor settings at certain loads.

During routine maintenance it was observed that the ignition timing had a significant effect on the power output of the engine. Judicious experimentation showed that advance beyond the normal setting for natural gas allowed greater power output. However, this caused starting problems both in cold starting with propane and sometimes on switch-over to landfill gas.

Eventually in 1986 the stage was reached where the excessive demand for power casued by the wear in the compressor feeding the brick kiln, overloaded the engine and shut down the system. The variation in demand for gas on the kiln was never reduced sufficiently to match the rather slow response time of the engine to load changes. After repeated spurious shut down events, the generator was taken out of service.

Although Technically unsatisfactory, the work had shown the Board of Shanks & McEwan (Southern) that landfill gas could be used to generate electricity and hence, allowed progression to the next stage of electricity generation.

STAGE II GENERATION

The landfill site at Stewartby was a clay pit used by London Brick Company between 1947 and 1963. The electrical sub-station dated from 1946 and all the switchgear in it was old and very near the end of its useful life. Consequently it was not possible to take any additional supplies for the next phase of landfill gas control. In November 1985, an abstraction unit was commissioned using a diesel powered generator to supply power to the compressor, the after cooler and the instrumentation. The cost of this power was high and would have risen further with the planned expansion of gas abstraction. This lead to the design of a new electricity generating unit. A number of basic gas powered units were studied and a major concern became the degree of expertise of gas experience available as well as the service back up. Table 1 shows the short list of manufacturers deemed to be capable of supplying an installation with the basic costs of the prime mover with generator.

INSTALLATION

Although not the cheapest alternative, the Dorman Diesel 12STCWG was chosen. Table 2 gives the major characteristics. The installation was set out to have three 275kW units, leaving room for a fourth machine. The building was designed in-house, generator panels provided by the engine manufacturer and

the sub-station was built by Yorkshire Switchgear, incorporating a ring main unit on the 6.6kV side and a double sized 415 volt distribution board for local electrical supplies. This allowed the old clay pit sub-station to be dismantled and removed with modern switchgear substituting the obsolete units on existing electrical supplies. The capital justification was based both on the potential sales of electricity and the savings to be made in diesel generator hire and foresaw a three year payback.

The Department of Energy's Energy Technology Support Unit had watched with interest the work being done with the Rolls Royce unit and co-operated with the new installation. It was included in their Energy Efficiency Demonstration Scheme and a grant of 25% was made towards the capital cost of the project. As a result, the scheme has had independent monitoring since its inception. Although the installation of three Dorman Diesel engines was made more permanent than that of the Rolls Royce, the basic machines are similar. The only real difference is that the Rolls Royce worked on a stand alone basis and the Dorman Diesel machines run in parallel with the public electricity supply.

The three Dorman machines are mounted on a baseplate with anti-vibration mountings, directly coupled to the generator and running at 1,000 rpm. The engines have integrated fan cooled radiators which simplified the initial civil work and cut down the commissioning lead time. Air is drawn through louvres at the back of the generator house, over the machines and pushed outside through the radiators. This keeps the engine room cool and helps to cool the engines themselves. Fuel supply is under pressure delivered by a sliding vane compressor via two 15 micron filters and double regulating valves. Each cylinder bank of the Vee-form engine has its own air filter, turbocharger, intercooler and Impco diaphragm-type carburettor. The load is controlled by butterfly valve throttles downstream from the carburettor. Because of the lower calorific value of the landfill gas in relation to natural gas (some 60%), the engines have been derated by some 19%. Exhaust is via a single horizontally mounted silencer, to the outside of the building, positioned vertically above the engine. Lubricating oil is supplied from a bulk storage tank and topped up continuously from a "day tank".

The units generate power at 415 volts and feed to a distribution board within the engine room. Each panel houses its own synchronising gear, governor, voltage control and power integrators as well as indicating instruments. Above the panel are the bus bars from which the power is fed to the new packaged sub-station via a separate circuit breaker. Amongst normal protection on the breaker is a N.E.I. Allen R.O.C.O.F. unit which senses system disturbances and will trip the main circuit breaker before the plant can try to generate to an "islanded" system. The output from the plant is fed onto the busbars on the low voltage side of the package sub-station. This is used as the distribution point for "house" supplies, for gas abstraction and compression equipment and other supplies previously connected to the old sub-station. Surplus power is transformed to 6.6kV and fed onto the London Brick Company 6.6kV distribution system. In cases of prolonged faults on the Area Board system, there is a safety interlock on the isolator on the low voltage side of the transformer which will allow the R.O.C.O.F. unit to be disabled and the whole plant run as an island.

Major equipment was ordered in July 1986, civil works began in October 1986 and the plant was scheduled for commissioning in January 1987, although bad weather caused a number of serious delays. The units were commissioned from 17th February, 1987 and formally handed over on 11th March, 1987. The on-site commissioning work had been reduced by testing the component parts at Dorman's Stafford works. However, a number of simple electrical faults had to be taken out of the system in the first few weeks. Time was also spent in adjusting the engines to run on landfill gas as opposed to the natural gas used in the works testing sequences. In addition, the compressor used to supply gas to the engines, failed on three separate occasions. One such event resulted from the event of the oil supply pump motor. The other stoppage occurred when slugs of water entered the compressor and shattered the moving blades. Subsequently the moisture removal system was improved and the experience gained in operating with this type of equipment has lead to trouble free running since then.

LUBRICATION

European experience had shown that much of the important part of the care of a gas engine, concerned the type of oil used. The three characteristics of Total Base Number (T.B.N.), Ash and Pressure are difficult to optimise and inevitably lead to a compromise. Initially, the oil manufacturer specified a C.R.I.30 which was similar to a number of oils used in continental installations. Simple alkalinity tests were performed every 100 hours to estimate the remaining basic character of the oil, and a full analysis was performed every 250 hours. The oil was changed after 500 hours and 1,000 hours by which time sufficient experience had been gleaned to reduce oil changes to 1,000 hour intervals. At the 5,000 hour maintenance interval, the recommended work includes the fitting of service exchange cylinder heads. Whilst this was under way, it was noticed that lacquering had occurred in some of the cylinder bores and on the pistons. In addition the exhaust valves and valve seats and the valve stem and the valve rocker faces showed high wear. Consultation between engine and oil manufacturers lead to a change in oil specification to N.G. 30M oil with an increase in the extreme pressure (E.P.) additive and a higher T.B.N. At the same time, stellite valve seat inserts and valves were fitted to two machines. Also the water operating temperature was increased from 70°C to 80°C, to help the action of the oil additives.

At the 10,000 hour maintenance period, the effect of the new oil was scrutinised. Better wear control had been achieved with the E.P. additives and there was reduced lacquering. The higher T.B.N. had made control easier with the changed oil. With a greater range of alkalinity, simple oil tests were able to monitor the condition of the oil with greater accuracy. After discussions with both oil and engine manufacturers, the oil change intervals were extended to 1,500 hour intervals and the 5,000 hour and 10,000 hour maintenance intervals increased to 6,000 hours and 12,000 hours respectively.

Regular oil samples were taken and analyses for trace contaminants monitored closely. Both copper and silicon showed higher than normal levels. Elevated levels of copper in oil usually means that bearing wear is taking place. However, the main bearings and big ends are made from an aluminium and tin alloy and only the little ends made from phosphor bronze. The oil did not contain elevated levels of tin, so Dorman's concluded that the copper came from the oil cooler tubes. At first it was concluded that the silicon came from dust in the atmosphere around the adjacent brickworks. A special twin element air cleaner was fitted to one engine to try to reduce the levels. Another engine had an oil centrifuge fitted which operated on a bypass to the main oil filters and removed silica particles. Although in both cases, the silica levels had fallen, in neither case had they been reduced to more normal levels. It is possible that some silicon is leaking from the silicon rubber O-rings and sealants used in the engine but all engines have been fitted with twin element air cleaners and bypass oil centrifuges.

OTHER MODIFICATIONS

During the Summer of 1987, overheating and an alarm malfunction caused one of the generators to trip. This occurred in a very hot spell and lead to some more tests being undertaken into the optimum running conditions for the engines. Although it was found that the radiator core was blocked, there were higher temperatures also on the other two machines. Subsequently the radiator fans were increased in speed by fitting a larger pulley. Recent hot spells have not caused further trouble.

As mentioned earlier, the scheme is designed to be capable of running separately from the electricity board grid. At one stage it was intended that the original Rolls Royce should be used to start the system if "islanded" from the grid. However, Dorman Diesels showed that there was sufficient flexibility to use natural gas on the engines without modification. As a result, "black start" running is started using compressed natural gas from conventional cylinders. So far this has been kept only to periods during which the transformer or 6.6kV connection to the brickworks is being maintained.

NOISE

With the legislation represented by the Control of Substances Hazardous to Health (C.O.S.H.H.), noise is a fundamental problem. C.O.S.H.H. is more concerned by the effect of noise in the workplace but even low level noise has a residential nuisance value in the hours between 11pm and 7am. The building was designed as a cost effective anti-weather envelope but used inexpensive sound deadening materials. Hollow concrete blocks were used instead of bricks for the bottom two meters, double skinned insulated walls with steel sheeting on the outside were used above this and the roof was of similar construction with insulated skylights. There is a personnel door and an insulated 4 metre wide roll shutter door. All these leave a warm interior in winter but despite louvres for the air inlet to the building, a sound level outside the buidling measured at 1 metre, below 60dB (A). Within the building it is hard to control the noise without enclosing each engine in its own accoustic booth. Because oil over a period of time tends to soak into such structures and causes additional fire hazards, it was decided not to enclose the engines. The person taking daily readings wears ear muffs for his ¼ hour per day visit and fitters wear the same during routine maintenance.

TARIFFS

The adjacent brickworks runs for 80 hours each week. It has a maximum demand of some 5.6 MVA with the overnight and weekend loads dropping below the total output of the three engines. At such times the power is exported to the local electricity board grid which runs at 33kV. The approximate distribution of the use of power is:-

Internal Consumption	13%
Export to London Brick Company	77%
Export to Electricity Board	10%

Because the brickworks runs during the periods of high Winter rates, those high figures shown in Table 3 do not apply. The overall average rate (for export over 8,760 hours per year) is 2.75p per unit whereas the removal of the high figures gives a net annual average of 2.36p per unit. It is interesting to note the advent of the Privatisation of the Electricity Supply Industry promises better rates. However, the average rate being quoted by the Area Boards for future contracts is of the order of 2.8p per unit. The Non-Fossil Fuel Obligation should allow generators of electricity from landfill gas to make use of the scheme and it is unfortunate that no information on typical rates is available at the time of writing this paper.

CONTROL

The units are designed to run unattended for 24 hours each day and 365 days each year. A comprehensive system of alarms and trips allow the system to generate an alarm, and also to shut down the units in the event of difficulties such as low oil pressure, low gas pressure, high engine temperature or engine overspeed. These alarms also are connected by a telephone link to a central computer which monitors the installation (amongst others). This means that with a 24 hour call out system all emergencies can be met. Additionally the building is equipped with smoke and gas detection sensors which can shut down all the engines and also activate the 24 hour call out system.

FUTURE

At the third anniversary of the commissioning of the engines, they had run for over 75,000 hours and generated more than 20 million units of electricity. The success of the scheme lead to the installation of a fourth unit which has generated for over 6,000 hours already. It can be understood from this that electrical generation from landfill gas using spark ignition engines is firmly entrenched as one of the favoured paths of energy conversion within Shanks & McEwan. Several other projects are in the design phase and wait more for the outcome of the size of the non fuel fossil subsidy than any other reason. The machines in prospect range from similar sizes to an order of magnitude greater in size. Inevitably the largest sizes will be compression ignition dual fuel engines rather than spark ignition models. At that size (over 2MW) the extra efficiency of conversion of energy makes the more expensive machine more cost effective.

TABLE 1 - EQUIPMENT CONSIDERED FOR GENERATION

MAKER	UNIT SIZE	£/kW
Caterpillar - G399	500kW	220
Dorman Diesels - 12STCWG	275kW	310
N.E.I.-Allen - 6S37-E (Dual Fuel)	800kW	410
Ruston Diesels - 6RK270	800kW	470
Waukesha - VHP9500GSI	800kW	550

TABLE 2 - GENERATOR CHARACTERISTICS

Engine - Dorman Diesel - 12 cylinder Vee Form Turbocharged Water Cooled Engine

Bore	- 158.75mm (6.25 in)
Stroke	- 190.50mm (7.50 in)
Total Swept Volume	- 45.25 litre (2761 cu in)
Sump Oil Capacity	- 102.3 litre (22.4 gall)
Compression Ration	- 9.5:1
Speed	- 1,000 rpm

Impco Carburettor
Altronic Ignition System
Radiator Cooler with Fan Drive from Engine

Alternator - Newage Stanford

350kVA 415 volt brushless, self exciting, drip proof case.
Operating between 0.8 and unity power factor.
Two-thirds pole-pitch winding.

TABLE 3 – TARIFF QUOTED BY EASTERN ELECTRICITY.
EXPORT OF ELECTRICITY FOR YEAR FROM
1ST APRIL, 1987

PERIOD	RATE	HOURS/YEAR IN PERIOD
00.30 to 07.30	1.64p/unit	2555
Moday to Friday 07.30 to 20.00		
December and January	7.25p/unit	554
November and February	4.25p/unit	518
All other times	2.66p/unit	5133
Average	2.75p/unit	8760

SCHEMATIC LAYOUT OF STEWARTBY LANDFILL GAS SYSTEM

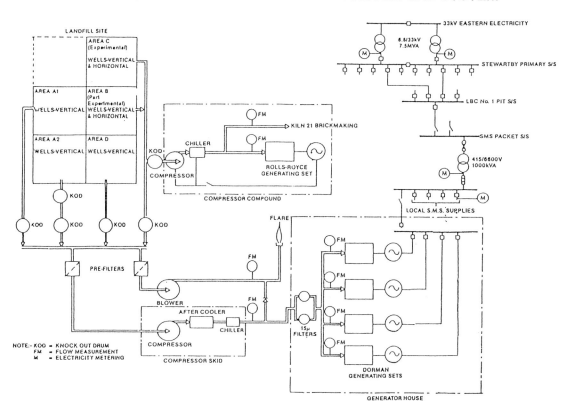

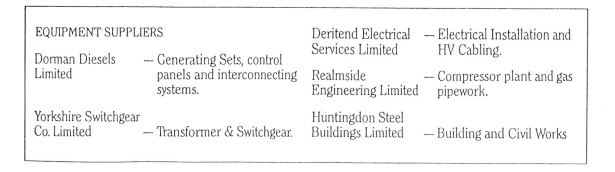

EQUIPMENT SUPPLIERS

Dorman Diesels Limited — Generating Sets, control panels and interconnecting systems.

Yorkshire Switchgear Co. Limited — Transformer & Switchgear.

Deritend Electrical Services Limited — Electrical Installation and HV Cabling.

Realmside Engineering Limited — Compressor plant and gas pipework.

Huntingdon Steel Buildings Limited — Building and Civil Works

Advanced packaged co-generation

J P PACKER, BSc(Hons), PhD
Director — Combined Power Systems Limited, Manchester, UK

SYNOPSIS Small-scale cogeneration, i.e. units of up to 500 kW of electrical output, is becoming of greater interest and is now being installed in a wide variety of applications. An advanced computer-controlled packaged cogeneration system is described together with a complete turnkey philosophy applied to the installation and support of small-scale cogeneration systems.

1. INTRODUCTION

1.1 Background

Interest in small scale cogeneration, defined as systems of up to 500 kW electrical output, was stimulated by the 1983 Energy Act (1). This Act simplified procedures for small generators to connect to— and run in parallel with the mains supply system (normally generating at 415V).

The vast majority of the units installed to date have been of skid-mounted construction with a typical system comprising

(a) reciprocating 4-stroke engine driving a 3-phase generator
(b) heat recovery system
(c) control and protection system

Fig 1 is a representation of a typical system.

1.2 Typical Systems

(a) Engines and Generators

(i) Engines: The majority of the units have operated on gaseous fuels (primarily natural gas) using spark ignition engines derived from compression-ignition engines. These converted engines are inherently reliable due to the lower mechanical and thermal loads experienced by the engines operating on gas compared to their diesel operation. With one notable exception the engines have operated at 1500 rpm (the exception being the Fiat "Totem" 15kW unit which used the Fiat 127 engine operating at 3000 rpm).

(ii) Two types of generator have been used viz:

(1) Synchronous: requiring engine control to synchronise the generator to the mains supply

(2) Asynchronous: these are in fact induction motors which are used to start the engine by connecting the motor to the mains supply.

(b) Heat Recovery System

There have been several variations of arrangement but whichever is employed the systems have transferred the necessary heat to cool the engine jacket and oil cooler, together with heat recovered from the exhaust gases, to the secondary water flow.

The majority of secondary water systems to which these units have been interfaced have been low pressure hot water (LPHW) systems with flow temperatures in the range 70 deg C to 90 deg C. In some installations (eg leisure centres) there is also a lower temperature heat load (eg pool water). As this circuit is typically at around 30 deg C it provides the means to extract further heat from the exhaust gases by condensing the water vapour in the exhaust (thus gaining the latest heat released during condensing).

With certain engines it is possible to operate at higher than normal cooling temperatures and thus interface to medium pressure hot water circuits (MPHW) with flow temperatures in the range 110 to 130 deg C.

At the larger unit size (ie approaching 500 kW) it becomes economical to pass the engine exhaust gases to steam-raising waste heat boilers, however a use is normally still required for the heat extracted from the engine cooling circuit. (eg boiler feed/make-up water pre-heating).

(c) Control and protection System

This is in its most basic form the electrical interface with the external electrical system. Since cogeneration sets will normally be operating in parallel with the "mains supply" the control and protection system will have to satisfy the local electricity authorities requirements, including, as a minimum requirement, the "G59" regulations (2). Beyond "G59" the degree of mechanical and thermal protection and the sophistication of the control system depends upon the supplier since there are no specific regulations to comply with (however the Energy Technology Support Unit have recently produced a "Good Practice Guide" (3)).

1.3 Typical Performance and Applications

If full energy input (gross calorific value) is considered to be reresented by 100% then typical CHP outputs are as follows:

electricity 25 to 30%
heat output 50%
condensing heat output 10%

giving overall energy efficiencies in the range 75% to 90%. The principal economic attraction of CHP is the energy cost saving it can provide viz:

* marginal cost savings primarily through displacing high cost electricity normally purchased from the local electricity supplier.

* electrical maximum demand saving (the achievement of which depends heavily on continuous operation through the maximum demand periods).

* Reduction in inefficient boiler operation. The simple analysis below provides an indication of the scale of marginal cost savings:-

Direct energy cost saving

$$\text{energy cost before CHP} = EP_e + \frac{Q \, pfb}{b} \ (p/hr)$$

where E = electrical load (kw)

P_e = price of electricity (p/kWh)

Q = heat load (kW)

b = "boiler" efficiency

P_{fb} = price of boiler fuel (p/kWh)

$$\text{energy cost after CHP} = (E-E_{chp})P_e + \frac{(Q-Q_{chp}) \, P_{fb}}{b} + \frac{E_{chp} \, P_{fchp}}{g}$$

where E_{chp} = CHP electrical output (kW)

Q_{chp} = CHP heat output (kW)

g = CHP electrical generating efficiency

P_{fchp} = price of CHP fuel (p/hr)

direct savings =

$$E_{chp} \, P_e + \frac{Q_{chp} \, P_{fb}}{b} - \frac{E_{chp} \, P_{fchp}}{g}$$

The costs which must weigh against these benefits include

* capital cost of equipment

* installation costs

* maintenance costs

In addition the user may have to change electrical tariff, depending on the specific case this can either be a cost or benefit.

In general potential users will be looking for simple payback of from two to three years and typical applications include:

* leisure centres

* hotels

* hospitals

* small-scale district heating/blocks of flats

* sheltered housing

* "biogas" digestion plant (sewage treatment/ farms)

* industrial processes

* landfill gas sites (though more normally as generating sets rather than cogenerators).

In 1988 the Dept. of Energy published a study (4) which identified over 500 sites suitable for the installation of small-scale cogeneration by 1995. This represented an installed generation capacity of approximately 500 MW. Recent developments (technical, marketing and legislative) have at least doubled this conservative potential.

1.4 Typical Installation

Electrically the cogeneration unit is connected to run in parallel with the mains supply however combined cogeneration/island generator operation is becoming increasingly popular.

Thermally the unit may be connected in parallel to boiler plant (Fig 2(a)) or in series with it (Fig 2(b)). The choice of the method of connection is site specific. Fig 2(b) shows the inclusion of heat dumping plant used to reject heat when the digester heat load is satisfied. Site assessment begins with the determination of the available heat load since in simple terms if the fuel for the cogenerator has to be purchased then a use for the heat must be found (although there are subtle exceptions to this rule). Calculation of available heat load is based upon bill data, site visit to establish actual heat plant installation, pump duties etc. and in very rare cases, monitored data. Having established the heat load the size of cogeneration installation is optimised to minimise payback. From this the next step is to assess whether electrical export is likely and, if so, whether it can be economically justified. If electrical export is not economic then the scheme has to prevent it occurring (either by reduced installation size or power modulation in response to export).

2. ADVANCED PACKAGED CO-GENERATION

2.1 DESIGN PHILOSOPHY

Combined Power Systems Ltd (CPSL) was formed in 1984 with the initial intention of developing a cogeneration unit of 40kW electrical output. Market and technical surveys were conducted which revealed that

- consistent, reliable performance is essential with high electrical generation efficiency

- potential customers could not/should not contemplate maintaining the units

- appropriate sizing and careful installation design are essential

This led to the conclusion that the company should be cogeneration specialists developing expertise in the design, installation and maintenance of cogeneration systems. From this the company's philosophy evolved:

* To produce a cogeneration system an unobtrusive and reliable as a boiler but which unlike a boiler "generates" substantial energy cost savings.

* To provide the complete turnkey cogeneration service encompassing

- initial energy data analysis

- unit selction and installation design

- design, development, manufacture and testing of the units,

- physical installation (electrical and mechanical)

- commissioning and subsequent service support.

2.2 THE "INTELLIGENT" COGENERATION UNIT

In order to comply with the company's maintenance support philosphy it is obviously essential to know how the units are performing. This led to the developmenht of CPSL's own computer system which is fitted to each of its cogeneration units.

* These on-board computers

- monitor the health of the unit (logging data)

- control the engine (synchronisation, governing, power modulation)

- protect the unit thermally, electrically and mechanically

- communicate with a central maintenance computer if the unit has a problem.

* The central maintenance computer is used to

- be able to contact any unit at any time to "see" what its doing

- extract the logged data from the units and store it in an advanced data base (a complete history of any units is therefore available).

- stop/start units and modify elements of the on-board software

2.3 THE ON-BOARD COMPUTER

Performance and Health Monitoring

Parameters are continuously monitored to ensure that the unit is operating within set limits. Parameters are stored in memory at synchronisation; at regular intervals whilst on load and on tripping.

The parameters include:

- exhaust temperatures (all cylinders and across exhaust gas cooler)

- primary water temperatures

- secondary water temperatures

- miscellaneous temperatures (oil, enclosure ambient etc.)

- pressures (oil, gas, inlet manifold)

- flows (fuel, water)

- voltages, currents - power, frequency etc.

- throttle position, oil level etc.

- If any parameter exceeds the set 'trip limit' the computer shuts the unit down in a safe manner.

Protection and Control

- The on-board computer decides whether or not the unit is able to run, it then controls the engine to perform the following.

- engine starting

- synchronisation to mains

- power governing subsequent to synchronisation

- power modulation in response to external conditions

Communication

* There are currently in excess of 80 different reasons for the machine to trip

* Each trip has a number of 'reclosures' associated with it

* In the event of a trip for which the reclosures have not been exhausted the computer will wait for the fault to clear and will then restart the unit.

* If the unit trips and the number of reclosures has been exhausted then the on-board computer makes an 'SOS' call to the central maintenance computer via a modem and a conventional telephone line.

* The 'SOS' message comprises unit number and reason for calling

* The central computer phones back to extract logged data from the unit. This data is therefore almost immediately available for analysis by service personnel.

2.4 THE CENTRAL COMPUTER

The central computer system is based around SUN workstations.

The data retrieved from the units is stored on a sophisticated data base (INGRES) which means that historical data is available on line.

In addition to this expert system software is being developed to assist prediction of faults and to plan maintenance.

From the central computer it is possible to

- stop and start units
- extract logged data
- "view" units in real time
- load/modify software

2.5 CPSL's CO-GENERATION UNITS

At the time of writing CPSL has packaged units in the range 38kW (electrical) to 185kW (electrical) with units at 290kW and 560kW to be launched during 1990. These are listed in Table 1. At the other end of the power scale 10kW prototypes will be on field trial during 1990. All units make use of the computerised control, protection and communication system described above.

To date all the units supplied (at the time of writing in excess of 120 machines) have operated on gaseous fuel, primarily natural gas but with a significant number on digester gas. All units have used synchronous generators which have desirable electrical characteristics including simple adjustment of power factor via variation of excitation.

All units include a complete heat recovery system shown schematically in Fig 3. In this system the engine ("primary") water circulates through the engine jacket, oil cooler, water cooled exhaust manifold and exhaust gas cooler. All heat transfer to the secondary water takes place in a single heat exchanger. Primary and secondary water pumps are electric (engine-driven pumps are not used). This provides pump-run on on unit shut-down which removes thermal soak problems and also provides the means to pre-heat the engine (providing the secondary water is hot).

The engine oil supply system and engine water system are also integral to the unit.

Fig 4 shows a compact cogeneration package (CPSL Type 6Fg) of 75kW(e) output and making use of a Ford 7.8 litre gas engine. This compact arrangement minimises the overall size of the unit. Fig 5 shows CPSL's 12Mg unit (185kW(e)) built around an MAN 24 litre gas engine. This unit shows the heat recovery system built into a lower module with the engine-generator sub-system mounted on top. This arrangement, although not as space efficient as the "compact" 6F is far simpler to construct. For both units the control and protection system is mounted at the engine-end of the skid.

Fig 6 shows CPSL's on-board computer system (fitted to all units) mounted inside the door of a control panel. The system comprises 3 board viz:

* the CPU board (bottom right)
* the interface board (bottom left)
* power supply and relay board (top left)

Fig 7 shows the arrangement of the electrical control panel of a 12Mg unit.

Conclusions

Small-scale cogeneration is now becoming an accepted technology in certain market area eg

- hospitals
- leisure centres
- hotels
- sewage treatment works

and is proving that substantial energy cost savings are achievable given the correct total cogeneration package.

CPSL itself has recently moved to new premises in West Manchester and is currently producing 12 units per month (approx 1.2 MW of generation per month). The new premises will eventually be capable of building and testing up to 30 units per month.

The ever-present interest in reducing energy costs, the new and burgeoning concern for the environment together with interesting developments emerging from the privatisation of the electrical supply industry all point to an exciting future for small scale cogeneration.

References

1) 1983 Energy Act, HM Government

2) Electricity Council Engineering Recommendations G59

3) Good Practice Guide 1
    "Guidance Notes for the Implementation of Small Scale Packaged Combined Heat and Power" E.E.O., Dept. of Energy, 1989.

4) "Potential for Small Scale CHP in U.K. Public Commercial and Domestic Buildings" ETSU Market Study No. 8 1988

Table 1

CPSL's packaged co-generation units

UNIT TYPE	ELEC. OUTPUT (kW)	HEAT OUTPUT (kW)	NAT. GAS CONSUMPTION (kW)
4Fg	38	70	135
6Fg	75	130	275
6Mg	95	150	310
12Mg	185	320	610
12MTg	290	420	950

4Fg   = 4 cylinder Ford engine
6Fg   = 6 cylinder Ford engine
6Mg   = 6 cylinder MAN engine
12Mg   = 12 cylinder MAN engine
12MTg = 12 cylinder turbo-charged MAN engine

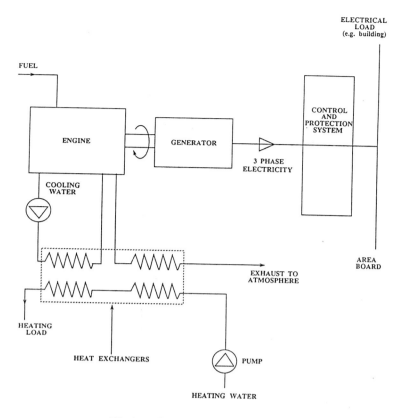

Fig 1   Co-generation schematic

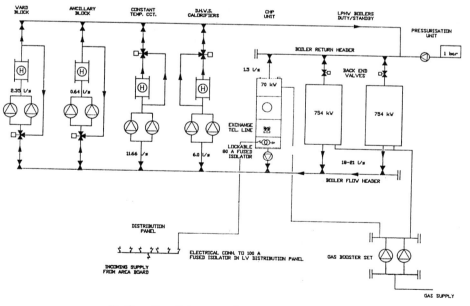

Fig 2a   Parallel connection to boiler plant

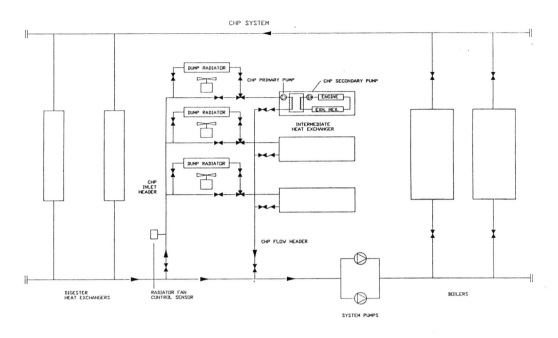

Fig 2b   Series connection to boiler plant

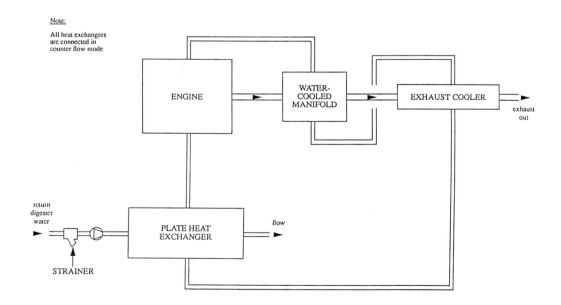

Fig 3   Heat recovery system

Fig 4   75 kW co-generation unit

Fig 5   185 kW co-generation unit

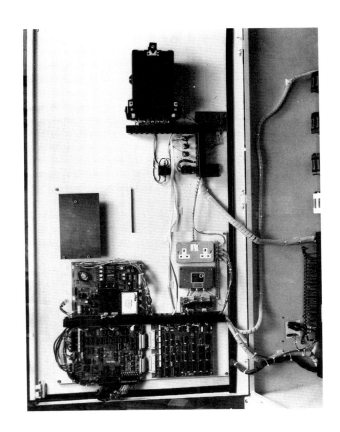

Fig 6   The CPSL on-board computer

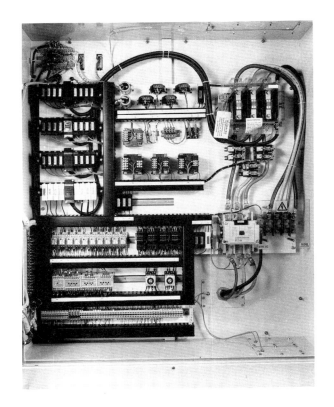

Fig 7   Control panel electrical arrangement

RK270 spark ignited engine

M WHATTAM, BEng, CEng, MIMechE and **S K SINHA**, BSc
Ruston Diesels Limited, Newton-le-Willows, Merseyside, UK

Synopsis

Over the past decade, Ruston Diesels has moved on from well established dual fuel engines of 254 and 318 mm bore, to the design and development of a range of 270 mm bore spark ignited gas engines based on their successful RK270 diesel engines. These engines utilise the concept of 'lean burn combustion' in order to ensure low exhaust gas emissions and good fuel economy. The company has considerable experience of running these engines in the UK, Far East and the United States for both power generation and mechanical drive. The paper follows and details the development and improvement of these engines up to and beyond the present day, and can be summarised in the following stages.

   Stage 1. Initial engines with two valve cylinder head, open bowl piston and twin spark plug ignition system, non lean burn combustion.

   Stage 2. Move to a lean burn concept by fitting a side mounted pre-combustion chamber or a pre-combustion chamber integral with the cylinder head. to improve exhaust emissions and fuel economy instead of alternative rich burn and catalytic converter system.

   Stage 3. Refinement of lean burn combustion with 4 valve cylinder head and centrally mounted pre-combustion chamber to give increased output and speed capability together with lower emissions and fuel economy. Combustion process investigation with in-cylinder gas sampling to give better understanding of combustion and emission formation.

   Development of microprocessor based electronic engine management system, thereby allowing automatic adjustment of the engine to accommodate varying site conditions and fuel types, while operating at optimum efficiency close to the knock limit.

1. Introduction

Large, gas burning, turbocharged engines have been in existence for many years in the form of dual fuel engines which require a small quantity of diesel fuel to provide ignition of the larger gas/air mixture. There was a need to develop an engine which would operate solely on gas for applications where diesel fuel was not readily available. Some years ago, Ruston Diesels began development of such an engine based on their successful RK270 diesel engine. The initial development involved no redesign of major components due to the lower firing pressures involved in gas operation compared with that of diesel and concentrated primarily on the cylinder head, gas feed arrangement and spark ignition. A brief summary of the first two phases of development are described along with the more recent third stage which is the subject of this paper.

2. Engine Specification and Performance Target.

Medium speed, turbocharged, intercooled engine. 6 and 8 cylinder inline, 12 and 16 cylinder 'V'.

Bore   :   270 mm
Stroke   :   305 mm
Performance :   14 bar bmep
Speed Range :   500 - 1000 r/min

2.1 Stage 1

The initial development was based around a two valve cylinder head. The choice of this head was made on the basis of the anticipated airflow requirements compared to that of the RK270 operating on diesel fuel, and the ease with which the layout afforded the fitting and accessibility of one or more spark plugs. The anticipated market in North America also provided a gas supply at high pressure, allowing a higher boost pressure and thus permitting sufficient mixture to pass through the single inlet valve for the expected power output (1).

   The combustion chamber shape was formed by a spherical bowl piston and recessed cylinder head, Fig 1, providing minimum flame travel for a given volume. Ignition was by means of a spark plug placed near the exhaust valve to reduce the occurrence of detonation, with a second spark plug mounted in the side of the head as an aid to starting. This second plug was grounded during normal operation.

   The performance achieved with the above configuration is shown in Table 1. The performance of the engine after Stage 1 of the development was acceptable for certain applications but improvement in fuel consumption was identified as an important requirement for other applications. Also at this time, environmental issues were beginning to surface and emissions were likely to be of considerable interest in the future, therefore a further stage of development set out to address these aspects of engine performance.

2.2 Stage 2

The automotive industry was already debating the problems of emission control and the concept of lean burn was adopted to provide a means of reducing the emission levels without seriously effecting the fuel consumption.

   The major problem concerned with the lean burn concept is that of stable ignition. The very weak mixture does not lend itself to good flame propagation in a quiescent combustion chamber. The solution to this problem lies in the use of a pre-combustion chamber (PCC), containing a richer mixture than that of the main cylinder. Once the burning has started in the PCC, the rapid rise in pressure forces the flame into the main cylinder. Due to geometrical constraints, the PCC was mounted in the side of the cylinder head in the position previously occupied by the second spark plug, Fig 2. In this configuration, the original plug was used as a starting aid but grounded during normal operation. An alternative arrangement involved a cylinder head with an integral PCC. Both these designs have operated successfully in service.

Table 1 shows the improvement in performance gained during stage two, with above 20% improvement in fuel consumption. Stage two also confirmed that for low emission levels, the air/fuel ratios required were such that a lean burn/PCC arrangement was essential.

The performance at 800 r/min met commercial requirements, however a need to increase the power output and the engine speed to 900 and 1000 r/min resulted in a further development stage. The engine build at that time was not suited to the higher engine speeds due to the gas exchange requirements.

2.3 Stage 3

It was decided to return to a four valve cylinder head, but with a completely different design from that of the diesel variant. A number of research and development projects were undertaken to optimise the cylinder head from the point of view of airflow, gas flow and exhaust outlet. In the process of this redesign, it was appropriate to reposition the PCC centrally between the four valves, Fig 3. The gas input valve was also moved from a position just above the inlet valve head to a position further upstream, close to the inlet manifold. This encourages better mixing of air and gas before entering the combustion chamber.

The following discussion is concerned with the development of the four valve head with centrally mounted pre- combustion chamber. There were two main objectives, optimisation of performance at 1000 r/min with particular reference to fuel consumption and emissions along with the need for better understanding of the processes involved in the interaction of the pre-combustion chamber and the main cylinder in order to help predict the effect of future changes on emission levels. The effects of changes in engine build and operating parameters are described below.

3. Results

3.1 Standard performance

Table 2. details the standard performance of the engine with the 1000 r/min turbocharger build at both 1000 r/min and 750 r/min. The fuel consumption at 750 r/min is significantly worse than that at 1000 r/min indicating that there is considerable scope for optimising the turbocharger for 750 r/min operation.

3.2 Exhaust wastegate control

Table 3. shows the effect of wastegate control with the 1000 r/min build at 750 r/min. The use of a smaller turbocharger nozzle and the wastegate in a mid position, column 2, gives almost identical performance to that of column 1 where the wastegate is closed. The wastegate is therefore an important feature in variable speed applications and should allow good fuel consumption over a wide load/speed range and also allows for operation under extremes of ambient conditions.

3.3 Valve Timing

Table 4 shows the effect of advancing and retarding the valve timing by rotating the whole camshaft. A full investigation of the effect of individual valve events has been presented previously, (2), and this change serves as an example of an increase in compression ratio without the need to change the piston crown. It was expected that the

previously observed trends of trade off between bsfc and NOx and the relationship between exhaust temperature and NOx would be apparent, however advancing the valve timing showed significant deterioration in bsfc and emission levels and retarding produced an increase in emissions despite a reduction in exhaust temperatures. Considering the cycle temperatures, Fig 4 helps to explain the significant increase in NOx due to higher cycle temperature near top dead centre, however more investigation is required before accurate prediction of NOx levels is possible through analysis of the cycle temperatures.

3.4 Charge Air Temperature

Table 5 shows the effect of charge air temperature controlled by the charge cooler water temperature. The exhaust temperatures are understandably higher with the increased air temperature due to an increase in charge cooler water temperature from 28C to 41C, but the most noticeable effect is the NOx emission which is approximately doubled.

3.5 Pre-combustion Chamber (PCC) Nozzle Configuration

The PCC serves to provide a high energy ignition source to the main combustion chamber which contains a very lean mixture. There are two aspects of the PCC arrangement which could be expected to strongly influence engine performance. Firstly, the geometry of the PCC, in particular the number and size of holes providing jets of burning gas into the cylinder and secondly the mixture within the PCC which must be ignited in the conventional way using a spark plug.

A number of nozzle arrangements were investigated ranging from one to four holes of various sizes. Table 6 shows that the NOx emissions are governed primarily by the total hole area rather than the number of holes with a secondary consideration that for a given hole area, the single hole arrangement appears better.

In view of the simplicity of the single hole nozzle design compared with the more complex multihole designs and their potential manufacturing and life problems, the single hole nozzle is clearly superior.

Figs 5 and 6 show the effect of nozzle arrangement on emissions and fuel consumption, with the four hole nozzle giving slightly improved fuel consumption at the expense of a large increase in emissions.

3.6 PCC Gas Pressure

The mixture strength in the PCC is controlled by the gas pressure in the rail feeding the PCC. The effect of varying this gas pressure is shown in Table 7 and Fig 7. The effect on bsfc is relatively small although with a possible trend towards higher gas pressures yielding the lower fuel consumption. There is a point, when the mixture becomes too weak and bsfc increases more significantly. It would seem reasonable that a minimum gas pressure exists below which the PCC fails to ignite consistently. The effect on NOx emissions is less clear. The results show some scatter, of the order of ten per cent, and there is no clear trend.

3.7 Gas Sampling Valve

In order to further investigate the effects of the PCC and other parameters particularly with respect to the formation of NOx, it is important to consider the combustion process in more detail. Tests using an in-cylinder sampling technique developed by UMIST are underway and it is hoped that the results will shed some light on the apparent trade-off between fuel consumption and emissions. The sampling involves the use of a probe entering through the liner flange to make a controlled traverse across the combustion chamber. This probe takes a small sample of the gas in the cylinder in a very short space of time, building a picture of the in-cylinder conditions with respect to space and time. In addition to the probe in the cylinder, a similar probe is mounted in the exhaust outlet. This allows investigation of the formation of NOx after the main burning is complete.

The method of emission measurement takes into account the levels of CO_2, NOx, CO, SOx, CH_4 and other hydrocarbons in the exhaust system which is sufficient to provide information from the point of view of pollution.

3.8 Ignition timing

Figs 8-10 show the relationship between fuel consumption and NOx for different ignition timing while all other parameters are constant, at both 750 r/min and 1000 r/min. While specific fuel consumption reduces at higher loads, the NOx levels increase substantially. It is clear that there is a trade off between NOx and bsfc which can be controlled by ignition timing.

4. General

Comparing the results with those of stages 1 and 2, the four valve cylinder head design has allowed performance at the higher speeds and in fact gives some improvement in the trade off between fuel consumption and emissions. The original design target of 14 bar at 1000 r/min has now been achieved following stage 3 of the development.

5. Microprocessor control

From the preceding results, it is clear that engine performance, in particular emission levels, is sensitive to operating parameters. There is therefore much to be gained by dynamic control of these parameters.

In the past, control has been accomplished by means of a hybrid pneumatic system. The degree of control is restricted to defined limits since it relies on the interaction of relays, regulators, ratio devices etc. to provide relative positioning of the air butterfly valve, exhaust wastegate and ignition timing servos coupled with various pressure signals to the pre-combustion chamber and main gas supply regulators. Override control of the air butterfly and exhaust wastegate is applied once the exhaust gas temperature at turbocharger inlet exceeds a maximum value. The control curves used for the various devices are straight line relationships in order to provide a workable compromise.

The pneumatic system controls the air/fuel ratio by governing the gas pressure according to boost pressure ratio. Due to the modulation of the main gas pressure, the speed governor (actuator) reverts to a common output position after a load transient or speed change, once the engine has returned to a steady state condition. This method of control causes some difficulties concerned with overfuelling during startup and transient load changes with the governor tending towards a maximum fuel condition. A more refined control of gas/air mixture and ignition timing is clearly required.

In order to provide a more sensitive and flexible control system, Ruston has adopted the Regulateurs Europa Viking 22 system.

This microprocessor control system comprises :

An engine driven hydraulic actuator with electrically controlled pilot valve.

A microprocessor based controller.

Current to pneumatic transducers to interface with the existing servos.

The Viking 22 controller is programmed to carry out the following basic functions :

Control of gas supply to the engine via the hydraulic actuator.

Control of air supply to the engine using the existing air butterfly and exhaust wastegate according to engine load.

Automatic override functions are provided for maximum allowable turbocharger inlet temperature and air boost pressure.

Control of pre-combustion chamber gas pressure as a function of the air boost pressure.

Setting of the ignition timing as a function of engine load and speed.

Monitoring of fault shutdown conditions.

Start sequencing including lubricating oil priming, air purge, start air control, starting fuel limiting and associated alarms. Engine stop control is also included.

Start of auxiliary plant.

Normal speed governor functions including raise/lower load and speed, perception head failure detection, and gas limiting during transient load changes.

In the case of an engine/generator installation, the load is sensed via a KW transducer. In the case of a compressor or pump application, the control is based on speed sensing and load is determined from a user defined speed/load curve.

Previously governing was achieved by varying the gas pressure according to boost pressure, the digital controller allows for control of the gas input valve relative to load and therefore gives a proportional governing system, furthermore the gas pressure can be varied if required.

The starting and load transient overfuelling difficulties can be easily overcome by setting a fuel limiting curve based on boost pressure.

The most significant advantage of the microprocessor controller is the ability to adjust each control parameter independently. Fig 11 shows the control scheme as outlined above.

6. Experience

To date Ruston has accumulated over 140,000 hours of successful operating experience, including one installation which has completed over 30,000 hours. The engines in service are either six cylinder inline or twelve cylinder 'V' units spread between the U.K., Far East and the United States and ranging from power generation to pump or compressor applications and operating on natural or landfill gas.

Some of the engines are operating under extreme conditions driving compressors in the Colorado gas fields with temperatures ranging from -32C to +34C and often with dust particles in the intake air. A higher rate of liner wear was found due to the dust, but this was overcome by introducing superior air filtration and a suitable treatment of the liner material.

Operation for power generation purposes introduce problems associated with transient load acceptance and no load stability. Control of the latter proved no problem with adjustment of the ignition timing and gas butterfly. After synchronisation, the settings were adjusted again for the load demanded. It is a difficult task to keep totally free from engine misfire which is necessary for power generation particularly when operating in parallel with a grid system.

Engines running on landfill gas at Burbank, Los Angeles at 12.8 bar, 720 r/min have 8,000 hours experience and are operating within the emission requirements of South Coast Air Quality Management District Regulation.

It is considered that the experience is sufficient to be able to undertake other long term commitments with spark ignition engines.

References

(1) The Ruston RK270 range of spark ignition engines. M. Whattam & S.K. Sinha. Institution of Diesel & Gas Turbine Engineers, Publication No. 435.

(2) Developing the second generation Ruston RK270 spark ignition engine for low emission and improved economy. M. Whattam, S.K. Sinha and G. Moylan. D79, Cimac 1987.

Acknowledgement

The authors wish to thank the management of Ruston Diesels Ltd. for permission to publish this paper and Regulateurs Europa for their collaborative support. They would also like thank their colleagues who have contributed to this work, in particular, Mr. A. Bretton, Mr. D. Johnson and Mr. J. Tomlinson.

Table 1 : Engine performance

Reference	Stage 1	Stage 2	Stage 3
Speed (r/min)	800	800	1000
Bmep (bar)	11.4	11.4	14.1
BSFC (kJ/kWh)	13242	10435	9930
BSAC (kg/kWh)	8.27	6.61	6.26
NOx (ppm @15% O_2)			450
Ignition Timing (°BTDC)	25	20	15
Wastegate Position			Mid
Air Intake Temp. (C)	12	16	22
Comp. Exit Temp. (C)	140	144	113
CC Water Temp. (C)	50	55	34
Cyl. Outlet Temp. (C)	580	575	492
Turb. Inlet Temp. (C)	626	592	581
Turb. Exit Temp. (C)	505	469	457
Boost Pressure Ratio	2.53	2.58	2.03
Gas Pressure (bar)	3.41	2.79	3.2
PCC Rail Pressure (bar)			2.5

Table 2 : Standard engine performance - 1000 rpm build

Speed (r/min)	1000	750
Bmep (bar)	14.1	12.8
BSFC (kJ/kWh)	9930	10373
BSAC (kg/kWh)	6.26	6.44
NOx (ppm @15% O_2)	450	496
Ignition Timing (°BTDC)	15	5
Wastegate Position	Mid	Closed
Air Intake Temp. (C)	22	18
Comp. Exit Temp. (C)	113	84
CC Water Temp. (C)	34	28
Cyl. Outlet Temp. (C)	492	457
Turb. Inlet Temp. (C)	581	575
Turb. Exit Temp. (C)	457	465
Boost Pressure Ratio	2.03	1.69
Gas Pressure (bar)	3.2	2.1
PCC Rail Pressure (bar)	2.5	2.1

Table 3 : Effect of wastegate control

Ref	1000 rpm Build	750 rpm Build
Speed (r/min)	750	750
Bmep (bar)	12.8	12.8
BSFC (kJ/kWh)	10352	10408
BSAC (kg/kWh)	6.58	6.63
NOx (ppm @ 15% O_2)	186	199
Ignition Timing (°BTDC)	5	5
Wastegate Position	Closed	Mid
Air Intake Temp. (C)	15	14
Comp. Exit Temp. (C)	83	83
CC Water Temp. (C)	28	28
Cyl. Outlet Temp. (C)	453	461
Turb. Inlet Temp. (C)	569	575
Turb. Exit Temp. (C)	458	458
Boost Pressure Ratio	1.75	1.78
Gas Pressure (bar)	2.1	2.2
PCC Rail Pressure (bar)	1.4	1.5

Table 5 : Effect of charge cooler water temperature

Reference	1000	1000
Speed (r/min)	1000	1000
Bmep (bar)	12.8	12.8
BSFC (kJ/kWh)	9895	10018
BSAC (kg/kWh)	5.80	5.65
NOx (ppm @ 15% O_2)	488	1036
Ignition Timing (°BTDC)	15	15
Wastegate Position	Open	Open
Air Intake Temp. (C)	17.5	20.8
Comp. Exit Temp. (C)	88	88
CC Water Temp. (C)	28	41
Cyl. Outlet Temp. (C)	509	519
Turb. Inlet Temp. (C)	597	611
Turb. Exit Temp. (C)	483	495
Boost Pressure Ratio	1.72	1.71
Gas Pressure (bar)	2.7	2.8
PCC Rail Pressure (bar)	1.9	1.9

Table 4 : Effect of valve timing

•Reference	-10	Datum	+ 10
• Speed (r/min)	1000	1000	1000
Bmep (bar)	12.8	12.8	12.8
• BSFC (kJ/kWh)	9968	9865	10068
BSAC (kg/kWh)	5.88	5.70	5.61
• NOx (ppm @ 15% O_2)	1147	899	2122
Ignition Timing (°BTDC)	15	15	15
Wastegate Position	Open	Open	Open
' Air Intake Temp. (C)	17	19	20
' Comp. Exit Temp. (C)	88	89	85
CC Water Temp. (C)	40	40	40
' Cyl. Outlet Temp. (C)	500	517	552
. Turb. Inlet Temp. (C)	592	607	648
' Turb. Exit Temp. (C)	478	491	532
Boost Pressure Ratio	1.74	1.74	1.68
Gas Pressure (bar)	2.7	2.7	2.8
PCC Rail Pressure (bar)	1.9	1.9	1.7

Table 6 : Effect of PCC nozzle arrangement

Nozzle	4x1/8	1x1/4	2x3/16	1x5/16	1x10mm
Nozzle Area (mm^2)	32	32	34	49	79
Speed (r/min)	1000	1000	1000	1000	1000
Bmep (bar)	12.8	12.8	12.8	12.8	12.8
BSFC (kJ/kWh)	9730	10100	10018	9865	9724
BSAC (kg/kWh)	5.37	5.74	5.65	5.70	5.69
NOx (ppm @ 15% O_2)	1923	1309	1035	899	991
Ignition Timing (°BTDC)	15	15	15	15	15
Wastegate Position	Open	Open	Open	Open	Open
Air Intake Temp. (C)	20.2	20.2	20.8	18.7	20.7
Comp. Exit Temp. (C)	82	92	88	89	87
CC Water Temp. (C)	28	40	41	41	41
Cyl. Outlet Temp. (C)	510	517	519	517	507
Turb. Inlet Temp. (C)	598	609	611	607	593
Turb. Exit Temp. (C)	487	492	495	491	481
Boost Pressure Ratio	1.63	1.75	1.71	1.74	1.69
Gas Pressure (bar)	2.7	2.9	2.8	2.7	2.5
PCC Rail Pressure (bar)	1.6	2.0	1.9	1.9	1.8

Table 7 : Effect of PCC rail pressure

Speed (r/min)	1000	1000	1000	1000	1000
Bmep (bar)	12.8	12.8	12.8	12.8	12.8
BSFC (kJ/kWh)	10115	9995	10034	9919	9907
BSAC (kg/kWh)	6.24	6.23	6.27	6.24	6.24
NOx (ppm @ 15% O_2)	413	326	367	401	438
Ignition Timing (°BTDC)	15	15	15	15	15
Wastegate Position	Mid	Mid	Mid	Mid	Mid
Air Intake Temp. (C)	21	22	21	21	22
Comp. Exit Temp. (C)	98	99	97	97	99
CC Water Temp. (C)	28	28	28	28	28
Cyl. Outlet Temp. (C)	506	497	492	486	483
Turb. Inlet Temp. (C)	598	586	577	571	567
Turb. Exit Temp. (C)	478	467	461	456	452
Boost Pressure Ratio	1.86	1.85	1.83	1.84	1.86
Gas Pressure (bar)	2.9	2.8	2.8	2.8	2.8
PCC Rail Pressure (bar)	1.1	1.6	2.1	2.5	3.0

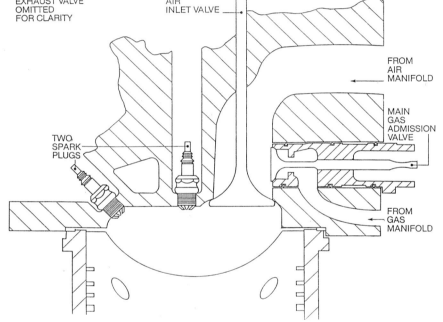

Fig 1   Stage 1 combustion chamber arrangement

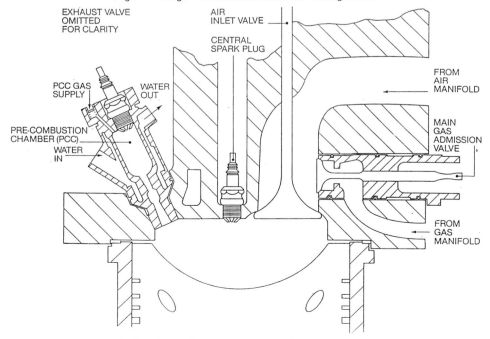

Fig 2   Stage 2 combustion chamber arrangement

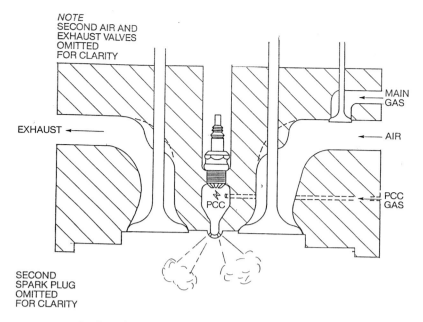

Fig 3   Stage 3 combustion chamber arrangement

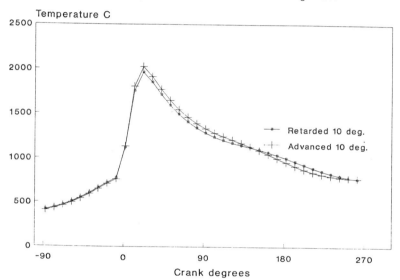

Fig 4   Effect of valve timing

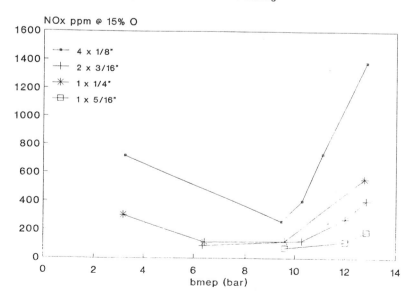

Fig 5   Effect of PCC nozzle arrangement on NO_x

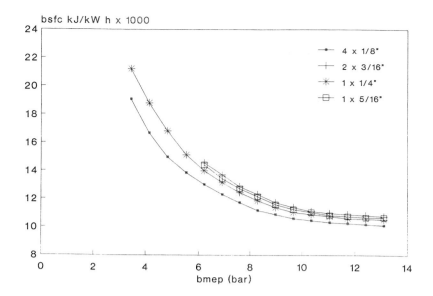

Fig 6   Effect of PCC nozzle arrangement on bsfc

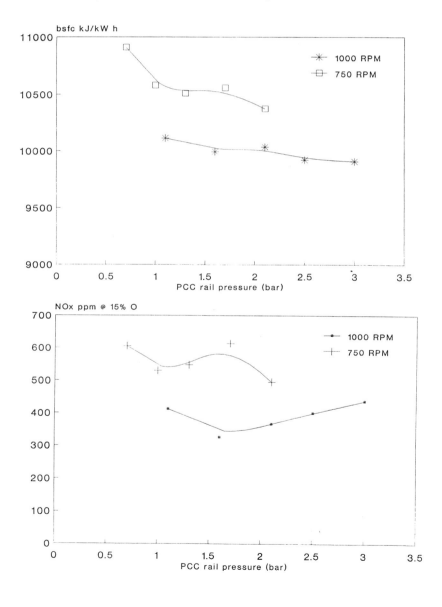

Fig 7   Effect of PCC gas pressure

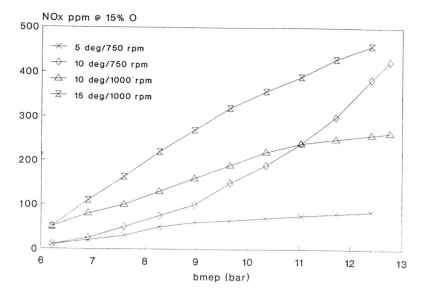

Fig 8   Effect of ignition timing on NO$_x$

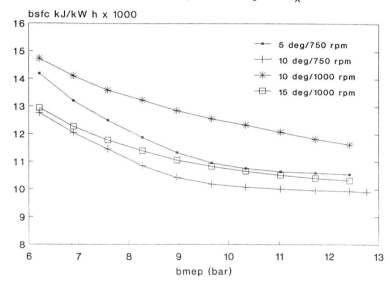

Fig 9   Effect of ignition timing on bsfc

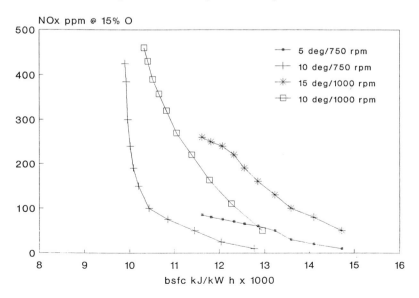

Fig 10   Effect of ignition timing, bsfc versus NO$_x$

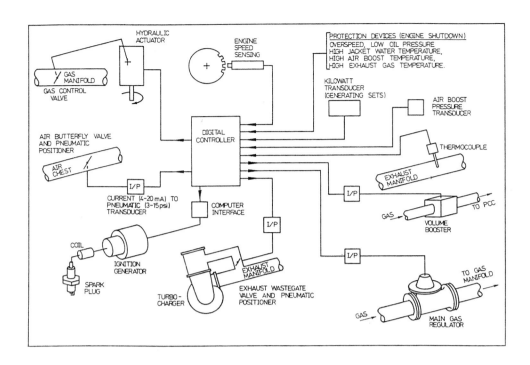

Fig 11   Microprocessor control scheme

The slow speed dual fuel engine for power generation

W J N WALKER, CEng, FIMechE
Kvaerner Kincaid Limited, Diesel Projects Department, Newcastle upon Tyne, UK
J O HELLMAN, BSCe
MAN B&W Diesel A/S, Copenhagen, Denmark

SYNOPSIS   The position and relative merits of the slow speed high pressure gas injection engine for power generation are reviewed. The high pressure system and other technical features of note are described.

1.   INTRODUCTION

The slow speed diesel generator is the most efficient, practical convertor of fossil fuel energy into electricity currently in operation. Its low fuel consumption, combined with high availability have made it eminently suitable for power generation and a recent survey has identified a total of over 400 two-stroke slow speed diesel engines installed worldwide for power generation. The stations range from those that have been operating for 40 or more years to new installations gaining full benefit from the latest technological advances associated with these engines. Overall thermal efficiency for slow speed diesel power stations can now exceed 50%.

A recent independent report (Reference 1) has found that for power stations in the range 20 - 150MW, out of all the options for fossil fuel burning plant, slow speed diesel engines are the most economical choice (see Fig. 1). The notation for this figure is as follows:-

GT   Gas Turbine
MSD   Medium Speed Diesel
OFS   Oil fired steam turbine
CFS   Coal fired steam turbine
CC   Combined cycle
SSD   Slow speed diesel
LFO   Light fuel oil
HFO   Heavy fuel oil

The recent order pattern has shown an increased awareness of the benefits to be obtained from this type of machinery and the successful adoption of high pressure gas injection technology makes these powerful machines available for gas burning power plants.

2.   BACKGROUND

The first engines to run using the high pressure gas injection concept were the two-stroke Nordberg engines produced in Milwaukee in the 1930's. More recently in the late 80's this principle has been adopted and tested by the principal slow speed diesel engine designers covering power outputs of between 1.5MW and 47MW from a single unit.

The engine operates on the two-stroke cycle with constant pressure turbocharging, employing the uniflow scavenge principle with a single, hydraulically operated exhaust valve in each cylinder.

The high-pressure gas injection system allows the same thermal efficiency and power output as a regular fuel oil burning engine. As the gas does not take part in the compression stroke, no risk of "knocking" exists, which in the past has been the limiting factor for mean effective pressure and hence power output of spark injected and dual fuel low pressure four-stroke engines. Also the engine with high pressure gas injection is independent of variations in gas quality and is able to use lean gases with a relatively low calorific value. The air to fuel ratio remains the same as for a fuel oil diesel, thus the heat load is the same and so no penalty in operation or maintenance results from burning gas. The result is a slow speed, dual-fuel engine with identical performance in terms of speed, output (no de-rating), thermal efficiency, exhaust gas amount and exhaust temperature as its well established heavy-fuel burning counterpart.

This ability to offer identical performance from a gas burning engine is an important step forward in the promotion of the dual fuel engine - see Fig. 2.

3.   THERMAL EFFICIENCY

An analysis of currently offered efficiencies for the various types of prime mover for power generation revealed the following:-

3.1 Low Pressure Gas Injection Engines

Low pressure gas injection engines cannot attain the same high efficiency as the high pressure G.I. (gas injection) engine and an efficiency of 42% at the alternator terminals may be taken as typical of current attainment. High pressure gas injection is now being applied to medium speed engines which might improve their position.

3.2 Gas Turbine and Combined Cycle

A product guide for modern large industrial gas turbines using manufacturers' data published in May, 1989 gave an average heat rate of 11,363 kJ/kWh from a list of 52 units quoted. This compares well with the figures of 11,386 kJ/kWh quoted for the gas turbines installed in a modern gas fired co-generation facility in Finland. This equates to a simple cycle efficiency of 31.6%. When running as an advanced combined cycle plant the station is quoted as having an overall heat rate of 7,709 kJ/kWh (corresponding to an efficiency of 46.7%).

3.3 Slow Speed High Pressure Gas Injection Engines

The slow speed diesel engine can be fitted with a turbo-compound system using excess exhaust gas energy to drive a turbogenerator to augment the power output from the diesel unit. Combined with energy optimisation of the engine this can result in efficiencies of up to 53% in terms of mechanical output. The engine can be offered in a format optimised for either power output where electrical output is the prime factor, or as an energy optimised version to satisfy a particular heat load profile. With the addition of a turbogenerator this can result in an overall efficiency for electricity production of some 50%.

Clearly the slow speed diesel can successfully compete with any plant in terms of efficiency and this is illustrated in Fig. 3.

A new gas turbine combined cycle plant being built in Turkey is claimed to be capable of returning a combined cycle heat rate of 7,008 kJ/kWeh (corresponding to an efficiency of 51.37%). However, this plant is still under construction and the design figures are yet to be realised in practice under site conditions. Such high efficiencies from modern advanced combined cycle plants, and the advent of STIG (steam injected gas turbine) pose both a threat and a challenge to the diesel designers.

4. PART LOAD OPERATION

Basic part load curves were constructed from information for a typical 33MW gas turbine and a 28MW slow speed diesel engine using typical full load heat rates. The basic figures were taken as relating to burning natural gas. Furthermore, it is possible that further losses amounting to some 2% penalty in heat rate could be accrued from inlet and exhaust losses for the gas turbine particularly in a combined cycle situation.

The resulting graph is Fig. 4 illustrating the flat curve for the slow speed diesel engine.

5. AVAILABILITY AND RELIABILITY

The slow speed diesel engine is able to consistently return high reliability and availability rates. Its fewer cylinders for a given output mean that even in a relatively unsophisticated environment, availability returns of 94% to 98% are being achieved. These figures are higher than for other prime movers despite the recent striking advances in availability being reported for gas turbine plant in the USA.

6. CO-GENERATION

The slow speed high pressure gas injection engine can be optimised to fulfil either an electrical power demand or a required heat output. Heat output optimisation is used for decentralised heat and power stations where the primary product is heat energy.

Fig. 5 shows the calculated possibilities for four different engine/plant configurations for a fixed electrical output of 20.0MW.

By comparison Fig. 6 shows the calculated possibilities for similar engine plant configurations for a fixed heat output of 17.7MW.

Actual operating installations for a baseload electrical plant utilising a turbo compound system and a bottoming cycle steam turbine, and for a co-generation facility at a pharmaceutical factory where the thermal energy of the exhaust gas and scavenge air is used in a fermentation process, demonstrate the possibilities from this type of engine – see Figs. 7 and 8.

7. THE HIGH PRESSURE GAS INJECTION ENGINE

Taking the basic oil-fuel burning engine with such inherent proven advantages as described above, the modifications necessary to safely adapt the engine to dual-fuel operation are confined essentially to the gas injection and control system on the engine.

Off-engine the station plant must be equipped with a suitably sized high pressure gas compressor capable of delivering the gas to the engine at a pressure of 250 bar and a temperature of between 40 and 50 Deg C. A buffer tank inclusive of condensate separator and a compressor control system are necessary. The appropriate safety devices and systems as described later in the paper must also be fitted to the engine and plant. Local regulations and standards to be complied with will have to be investigated with respect to each individual installation. However, as the engines have design approval from the major Clasification Societies such as Lloyds Register and Det Norske Veritas, these rules will be applied to the widest possible extent in the case of a lack of local regulations.

Technically the engine is identical to its standard fuel oil burning counterpart with the following exceptions ; gas supply pipework and distributor block with accumulator on a modified cylinder cover and a slightly modified camshaft system incorporating a control pump for the injected gas quantity. Minor modifications to the exhaust gas receiver, the exhaust valve and control system are also incorporated.

7.1   Gas Supply System

The high pressure gas from the compressor unit flows through the main pipe (positioned alongside the engine's upper gallery) via narrow and flexible branch pipes to each cylinder's gas distributor block and its associated buffer volume. The narrow and flexible branch pipes perform two important tasks:

-    They separate each cylinder unit from the rest in terms of gas dynamics, utilising the well-proven design philosophy of the standard engine's fuel oil system.

-    They act as flexible members between the stiff main pipe system and the engine structure, safeguarding against extra stresses in the main and branch pipes caused by the inevitable differences in thermal expansion of the gas pipe system and the engine structure.

Also the buffer volume, containing about 20 times the injection amount per stroke at MCR (maximum continuous rating), performs two important tasks:-

-    It supplies the amount of gas injected with only a slight, but predetermined pressure drop.

-    It forms an important part of the safety system.

Since the gas supply system is a common rail system, the gas injection valve must be controlled by another system, the control oil system. In principle, this is the fuel oil system of a standard small bore engine, consisting of its standard fuel injection pump, supplying high pressure control oil to the gas valve, thereby opening the gas valve and keeping it open during the period necessary for injecting the required amount of gas.

The normal fuel injection pump, which supplies pilot oil in dual fuel operation mode, is connected to the high pressure pipe of the control oil pump by a combined timing and safety valve.

The control system incorporates a balancing arm as well as controllable positioners in the two regulating shaft systems. By means of these positioners, the engine can be operated in the various relevant modes: normal dual-fuel mode with minimum pilot oil amount; fuel-oil-only mode and fixed-gas mode. The latter is only relevant in case of a limited gas supply from the external source, this amount being less than the engine needs to develop the required output.

7.2   Governor

The engine design is, as standard, prepared for electronic types of governors.

On the basis of the selected mode of operation, i.e. fuel-oil-only, dual fuel or fixed-gas mode, the positioners for the fuel oil and control oil racks will be controlled accordingly.

The governor is connected to a control box, which is connected to the safety system of the engine so as to exclude gas operation in case of relevant alarm. The control box also ensures that no gas can be injected below a pre-determined load via the load control.

The engine is started on fuel-oil-only with the gas setting positioner in zero. In this situation the gas pipe system is nitrogen-filled due to automatic shut-down of gas pressure followed by nitrogen purging when the engine load falls below 35 - 40%.

At loads above 35 - 40% the engine can be switched over to gas operation in dual fuel mode.

This is effected by gradually reducing the fuel oil amount by means of the fuel rack positioner to the minimum pilot fuel amount of 8% MCR fuel amount. Hence the control system will automatically increase the setting of the control oil pump so as to supply gas to the engine corresponding to the load controlled by the governor.

The same procedure is automatically followed in the reverse order when the load is again reduced to below 35 - 40% MCR, followed by automatic nitrogen purging of the gas fuel system.

A built-in memory in the control system prevents gas filling after nitrogen purging for a pre-determined period of 30-60 min. to prevent too frequent change-over, as well as to stop any change-over to gas without the engine being warm.

The automatic change-over from gas/fuel operation to fuel oil only will take place over an adjustable time period.

Alarm situations may lead to automatic cut-off of the gas supply followed by purging with nitrogen.

Provided this alarm is of a nature that, although cutting-off the gas supply, does not shut-down the engine, the governor will maintain the engine load, as the control oil positioner will increase the fuel oil amount accordingly. However, in order to avoid premature blocking of the gas injector passages during fuel oil operation, the reason for the gas shut down should be found and corrected at the first given opportunity.

7.3 Fuel Injection Valves

The injector for liquid and gaseous fuel constitutes a key component of the adaptation to dual-fuel burning.

The media to control and to be controlled by the injectors are gaseous fuel, liquid fuel, control oil and sealing oil.

Basically, liquid fuel is injected first and allowed to ignite as in a normal diesel engine after which the gaseous fuel is injected alongside the burning fuel, which acts as a pilot fuel for the gas. The gas injection is controlled by the control oil, and the sealing oil keeps the media apart in the injectors.

Based on comprehensive tests and modelling it was decided to design the injectors as two separately identifiable valves for liquid fuel and for gaseous fuel, but integrated into the same combined valve housing so as to ensure a close to parallel injection of liquid fuel and gas.

The two valves are fitted in a joint housing and share the same atomiser body, but the two parts can be controlled completely independently as

separate fuel oil and gas injection valves. The position and direction of the fuel oil sprays and the gas jets are similar, leading to the same thermal load pattern on the combustion chamber walls, and ensuring stable ignition of the gas in dual fuel operation with a minimum supply of pilot oil.

This valve type illustrated in Fig. 9 has been tested both in a 1000 h durability test in the workshop and in actual gas burning operation on an engine where one cylinder was converted to gas injection. Also valves have undergone durability testing, and have been endurance tested over a two year period with very good results.

The valves allow operation solely on liquid fuel up to at least 76% of MCR. If the customer's demand is for the gas engine to run at any time on 100 percent load with liquid fuel, without stopping the engine for changing the injection equipment, the fuel valve nozzle holes will be as the standard type for normal fuel oil operation.

The pilot oil can be heavy fuel and/or diesel oil, as preferred in each individual case.

Whereas the cylinder cover and the regular fuel pump are practically unchanged, the control oil systems comprise the fitting of an additional fuel pump on the camshaft housing. The control oil operating medium is the same as for pilot oil and, in the standard execution, forms an integrated part of the fuel oil circulating system.

For the sealing oil system, which is separate, a small high pressure pump, with a supply pressure higher than that of the high pressure gas system, is required. The consumption is less than 0.15 g/kWh, and heavy fuel or lubricating oil can be used.

Should long-term operation on 100% fuel oils become necessary for economic reasons during service, then the normal fuel injection valves can be retro-fitted into the cylinder cover and the engine operated as a standard oil burning diesel.

7.4 Gas Pipes

The gas pipes are double-wall piping, with the outer shielding designed so as to prevent gas outflow to the machinery spaces in the event of rupture. The intervening space, including also the space around valves, flanges etc., is equipped with separate mechanical ventilation with a capacity of at least 10 air or inert gas changes per hour. The pressure of the air/inert gas in the intervening space is kept below that of the machinery space, and fan motors are placed outside the ventilation ducts. The vent is connected to the chimney.

The gas pipes on the engine are designed for 375 bar pressure and are supported so as to avoid mechanical vibrations. The design is all welded as far as practicable, with flange connections only to the extent necessary for servicing purposes. The branch pipes are standard high pressure fuel injection pipes, which give impeccable service reliability at the 0-900 bar varying pressures - so the gas system pressure of some 250 bar ± 10 bar is no problem.

The gas pipes are connected to a nitrogen purging system, which is activated when the engine changes over to the fuel-oil-only mode.

7.5 Gas Distributor Block

An important new engine component is the gas distributor block. It has been found practical to incorporate all valves in a separate block per cylinder, bolted to the side of the cylinder cover and carrying the accumulator aswell. This provides a compact design and facilitates overhauling work on the engine. To pull a piston, for example, the cylinder cover with exhaust valve and gas distributor block is removed as a unit, the only extra work compared to an oil burning engine being removal of the two flexible gas branch pipes.

Except for the exhaust gas receiver being fitted with safety valves sufficiently dimensioned to prevent excessive pressures in the unlikely event of ignition failure of one cylinder followed by ignition of the unburned gas in the receiver, other engine modifications are limited to the already mentioned control system. Besides this, only for changed positions of pipes, platform cut-outs etc., will the gas injection engine differ from its oil burning counterpart.

8. SAFETY SYSTEMS

The normal safety systems incorporated in the fuel oil systems are fully retained for dual fuel operation. However, additional safety devices are incorporated in order to prevent situations which might lead to failures.

The key target has been to design for the prevention of faults rather than to detect faults when and if they occur.

In the engine hall of a power station, the presence of gas will be detected by a hydrocarbon analyser (HC). The same applies for the ventilation of the intervening space of the double wall piping, where alarm is given at a gas concentration of 30 percent of the lower explosion limit and shut down occurs at 60 per cent. Whereas all starts and

stops must be carried out on fuel oil only, the safety devices on the engine will continuously monitor the functioning of the components involved in gas operation during running.

When changing over from gas operation to fuel oil operation at a predetermined load, the entire gas pipe system will be purged with inert gas, e.g. nitrogen. When the load of the engine is decreased, the changing back to fuel oil will only take place automatically at the same predetermined load, and the gas pipes are again purged automatically by inert gas.

Possible main sources of faults during operation of the engine are:-

- Defective gas injection valves.

- Ignition failure of injected gas.

In order to prevent large amounts of gas from entering the engine following a possible defective gas injection valve (seizure and/or sluggishness), the gas flow to each cylinder is monitored.

If the gas flow considerably exceeds the value corresponding to the actual load, the pressure in the accumulator will drop below the corresponding shut down level, and the shut down valve in the gas distributor block will be activated.

Any abnormal gas flow, whether due to seized gas injection valves or fractured gas pipes, is detected immediately, the gas supply is discontinued, and the gas lines purged with inert gas.

Very small gas leakages are detected by exhaust gas temperature monitoring and by the hydrocarbon content in the exhaust gas, while large gas leakages are detected by a pressure drop in the accumulator.

Failing ignition of the injected natural gas can have a number of different causes, most of which, however are the result of failure to inject pilot oil in a cylinder.

To ensure that no gas is injected in the case of no pilot oil injection, the control oil pressure for gas valve opening is allowed only to reach the gas valve in the event of pilot oil pressure having built up. This is done by the pilot oil pressure activating a valve that otherwise will puncture the control oil circuit. One such valve is fitted in each cylinder's fuel injection system.

A second puncture valve for the control oil circuit is located on top of each exhaust valve so as to ensure that no gas injection takes place in the case of no compression and hence no ignition due to a sticking exhaust valve.

The safety systems are summarised in tabular form below in Appendix I. By adopting such measures it is possible to safeguard the engine installation and operating personnel.

9. ENVIRONMENTAL ASPECTS

Slow speed diesel power stations are more environmentally friendly than any other form of power generation using fossil fuel.

The principal environmental hazards currently receiving most attention and emanating from power stations are:-

Greenhouse Effect
Acid Rain
NOx

9.1 Greenhouse Effect

The greenhouse effect has recently attracted particular attention from environmentalists and the public. Some 45% of the greenhouse effect is said to arise from the carbon dioxide produced when fossil fuels are burnt. Because slow speed diesels burn less fuel per unit of electricity produced than any other form of conversion of fossil fuel energy, it follows that they therefore produce the least amount of carbon dioxide per unit of electricity and thus make the minimum contribution to the greenhouse effect, over their whole load range, combined with high efficiency. The CO emissions for gas turbines can actually increase at reduced load, particularly if steam/water injection is used for NOx control.

At the World Energy Conference in Montreal in September 1989, particular attention was paid to the environment and it was recognised that energy efficiency was the way to limit harmful emissions. Britain's Energy Secretary is reported to have said in his address to the Conference "energy efficiency is the single, most cost effective, physical response to the effort to limit carbon dioxide emissions".

9.2 Acid Rain

Sulphur dioxide is of course the main source of acid rain so it is no surprise that in Britain the CEGB are concentrating their efforts on SOx removal from their large thermal power stations.

Slow speed diesels can run on gas at the same high efficiency as they achieve on cheap residual fuel oil. The slow speed diesel engine's better efficiency means that it produces the

least amount of sulphur dioxide, fuel for fuel, than any other fossil fuel burner.

9.3 NOx

Acid rain is also caused by nitrous oxides and the CEGB have announced their intention to reduce the emission of nitrous oxides from their thermal stations by 30% by the end of the century.

Due to the high combustion temperature inherent in slow speed diesel engines, they produce higher than average levels of nitrous oxide emissions. However, by installing standard selective catalytic reduction (SCR) plant as part of the equipment for a slow speed diesel power station, the levels of nitrous oxide can be kept well within the most stringent limits in the world.

The slow speed diesel power station at Puerto Rico which comes under US legislation, Federal Rules of US EPA (Environmental Protection Agency), was the first such station to have nitrous oxide control techniques applied, and the results have proved to be consistently successful. For more stringent regulations SCR equipment is to be installed, and two engines with such equipment installed and complying with specific requirements have recently been delivered.

7.4 Noise

In addition to the above, noise can be an important consideration in the siting of a power station.

Typical noise levels at full load, 1m from the prime mover and in the exhaust are shown in Fig. 10. The higher values returned by the gas turbine reflect the greater attention to silencing and acoustic treatment of buildings or enclosures that must be applied when selecting a gas turbine.

The noise emitted by the slow speed diesel does not produce the same annoyance as the gas turbine when close to a residential area. There are slow speed diesel power stations sited harmoniously in urban housing areas, and this is obviously important if district heating is foreseen as an outlet for the utilisation of the heat energy.

SUMMARY

The advantages of the slow speed high pressure dual fuel engine for power generation can be listed as follows:-

1. High thermal efficiencies and thus low fuel consumption.

2. High reliability and availability.

3. Identical output and performance as the standard fuel oil version of the engine.

4. Flat heat rate curve at part loads.

5. Environmental impact low.

6. Possibility of simple revision to oil burning and vice versa should economic factors dictate.

REFERENCES

1. EPD Consultants Limited Generating plant options for power stations, October, 1989

ACKNOWLEDGEMENTS

The authors would like to thank the management of MAN B&W Diesel A/S and Kvaerner Kincaid for permission to publish this paper, and their colleagues for the contributions to its preparation.

APPENDIX I

Table 2   Engine and exhaust system mandatory safety devices

Fault	Safety devices
Defective Gas Injection Valve	
Sluggish operation	Exhaust gas temperature sensor, giving alarm for high temperature on the cylinder concerned.
Moderate gas leakage to combustion chamber	Exhaust gas temperature sensor, giving alarm for high temperature on the cylinder concerned.
Large gas leakage or seizure of gas valve in open position	Pressure controlled safety switch in the gas pipe will give immediate shut-down for gas supply and evacuate the injection system for gas. Safety valves on exhaust gas receiver will prevent damage from explosions in the exhaust gas receiver.
Ignition Failure	
Missing pilot oil injection	Safety valve in the pilot oil fuel injection pumps prevents gas injection.
Slight sticking of exhaust valve	Exhaust gas temperature sensor, giving alarm for low exhaust gas temperature on the cylinder concerned. Safety valve on exhaust gas receiver will prevent damage from explosions in the exhaust gas receiver.
Severe sticking of exhaust valve	Safety valve in the exhaust valve housing prevents gas injection.
Knocking	
One fuel oil valve sticking in open position, so that only this valve injects pilot oil.	The safety valve in the pilot fuel injection pumps prevents gas injection.
Severe gas valve leakage into the cylinder	Exhaust gas temperature sensor gives alarm for high exhaust gas temperature on the cylinder concerned.

Table 1   Engine room mandatory safety devices

Fault	Safety devices
Fractured gas pipe Leaky joints in gas pipes Leaky valves in gas pipes	Double-wall piping, covering all pipes, flanges and valves, with ventilation of the intervening space. Ventilation air to be checked by HC-analyzer with alarm. Fractured gas pipes will also lead to shut-down for gas supply by means of a pressure controlled safety switch in the gas pipe. Gas pipes will subsequently be automatically purged with inert gas. Engine room ventilation ensuring good air condition in all spaces and excluding the possibility of formation of gas pockets in the upper parts of the engine room. The ventilation air to be checked by HC-analyzer with alarm.

Optional Safety Devices

Fault	Safety devices
Fractured gas pipes Leaky joints in gas pipes Leaky valves in gas pipes	Hoods on engine top and at main gas valve group with ventilation air to be checked by HC-analyzer with alarm.

Table 3   Engine and exhaust system optional safety devices

Fault	Safety devices
Defective Gas Injection Valve Sluggish operation and gas leakage to combustion chamber.	HC-analyzer, monitoring exhaust gas, with alarm for increasing HC-level.
Ignition Failure Missing pilot oil injection	Monitoring of cylinder peak pressure by means of strain gauges on cylinder cover studs.
Sticking exhaust valve	Computer controlled monitoring of cylinder pressure at 30 degrees CA before TDC by means of strain gauges on cylinder cover studs.
Knocking	Monitoring of high frequency cylinder pressure fluctuations by menas of computer processing of signals from strain gauges on the cylinder cover studs.

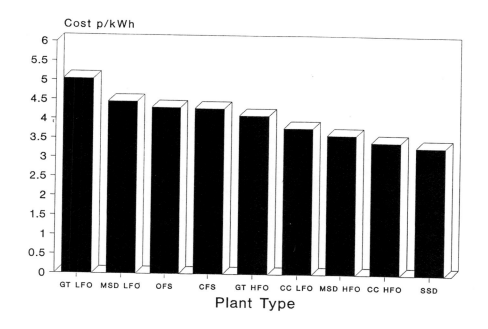

Fig 1   Alternative power generation proposals

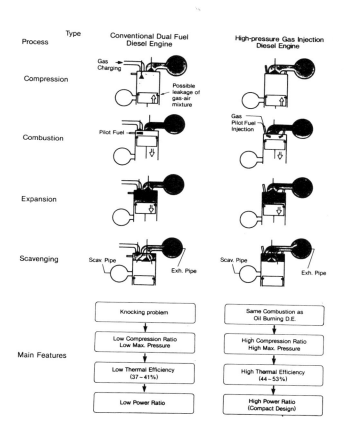

Fig 2   Dual fuel operation

	L.P. Gas Engine	Combined Cycle	H.P. Gas Engine
	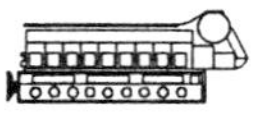 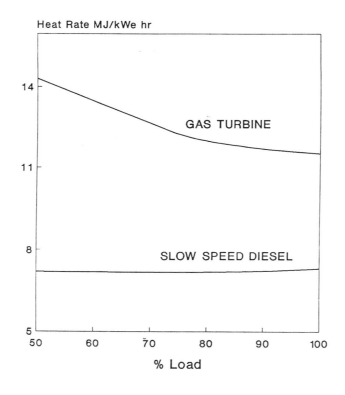	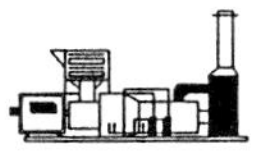	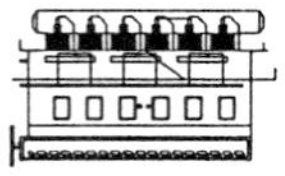
Overall Thermal Efficiency	41%	46%	48%
Availability	80%	90%	95%
* Gas Price	17.5p/Therm	17.5p/Therm	17.5p/Therm
Fuel Cost p/kWh	1.47	1.31	1.24
O&M Cost p/kWh	0.60	0.55	0.40
TOTAL GENERATING COST	2.07	1.86	1.64

* Based on interruptible Gas Tariff Band 5   (assumed gas LCV 48322kJ/kg)

Fig 3   Generation options

Heat Rate MJ/kWe hr

GAS TURBINE

SLOW SPEED DIESEL

% Load

Fig 4   Part load operation

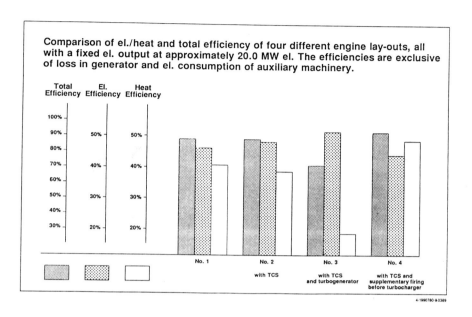

Fig 5   Heat balances for different plant configurations
optimised for electricity production

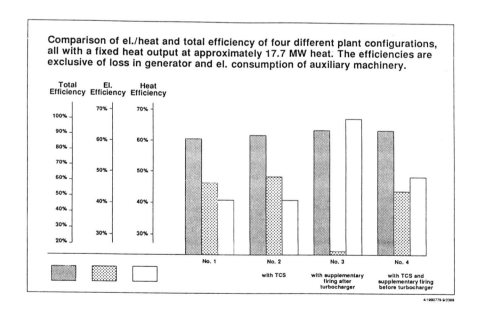

Fig 6   Heat balances for different plant configurations
optimised for heat production

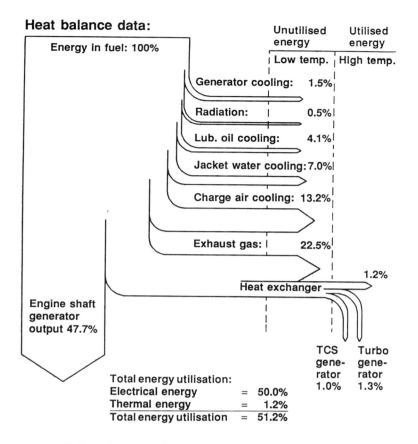

Heat balance data:

Fig 7   Power distribution for an actual baseload electrical plant

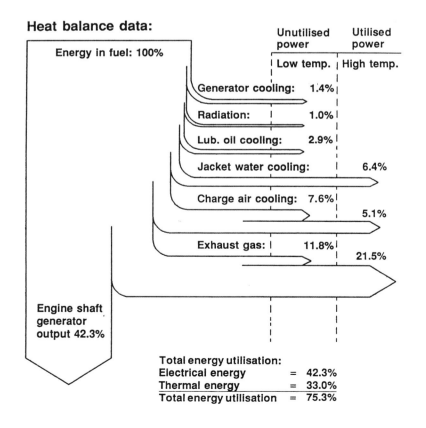

Heat balance data:

Fig 8   Power distribution for actual co-generation at a pharmaceutical factory

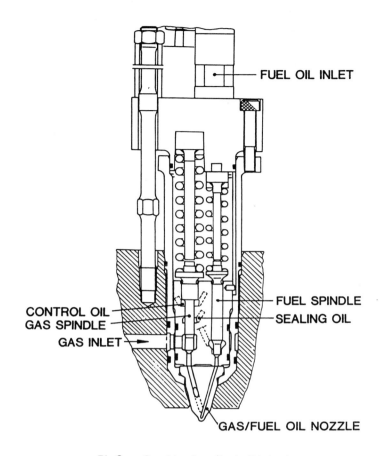

Fig 9   Combined gas/fuel oil injection valve

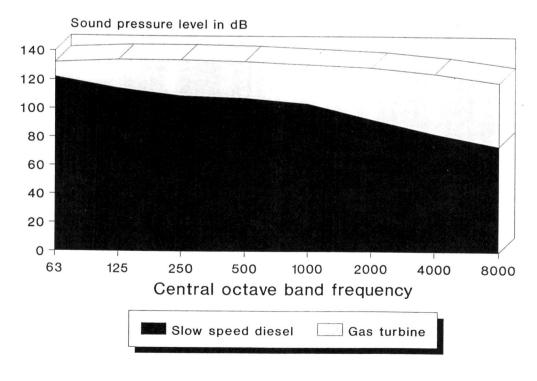

Fig 10   Average exhaust noise level

Wankel rotary gas engines

A W WINTERBOTHAM, CEng, MIMare, MRINA, **D EIERMANN**, Dipl-Ing and **R NUBER**, Dipl-Ing
Wankel GmbH, c/o Lonrho PLC, Cheapside House, London EC2, UK

SYNOPSIS

Interest worldwide in small engines designed to burn gas has been increasing in recent years due to growing awareness of the possibilities of making use of bio-gas and land-fill gas as alternatives to oil, in addition to the increasing availability of natural gas at reasonable cost.

Wankel GmbH in Lindau, West Germany, has developed a completely new range of gas burning engines using the Wankel Rotary Concept with significant success. The results show excellent efficiency and good durability. The innate suitability of the Wankel Rotary Engine for burning gas, along with the better known advantages of small size, low weight and smoothness, makes the Wankel engine a leading contender for the small scale gas engine market.

1. INTRODUCTION

Rotary engine technology was born in 1956 when the first gasoline Wankel rotary engine ran at Lindau, West Germany. The potential of this concept is still overshadowed by the immense inertia of piston engine technology, although recent advances in engine design and production method are rapidly shifting the balance in favour of the rotary engine in certain niche markets where its advantages can no longer be ignored.

The change in the rotary engine's fortunes are in many ways due to developments associated with piston engine technology. New manufacturing techniques and the availability of low cost production equipment make medium batch production of specialised engines an attractive option. Furthermore, the modern methods of controlling ignition and injection, originally developed for the piston engine, now offers the opportunity for rapid improvements of rotary engine performance.

Recent results have shown the potential for rotary engines to burn gas efficiently and reliably. The various characteristics of gas fuels are well suited to the features that the Wankel rotary engine exhibits and this matching provides a significant niche for the Wankel engine to fill.

2. FEATURES OF THE WANKEL ENGINE

The characteristics of the Wankel engine are well known, such as small size, low weight, smoothness, and excellent torque at low speed (1). How the Wankel rotary engine performs when burning gas is less widely understood. Rotary engines can be converted to burn most types of gas as easily as can piston engines. For example, the engines which were the subject of the durability tests undertaken by "Southwest Research Institute" (2) in 1985 and 1986 were adapted from stock Mazda 13B gasoline engines with virtually no adjustments to the main components. The modifications consisted of fitting a conventional gas mixer to the engine, adjustments to the ignition timing and re-routing the lubricant supply piping to take oil from an exterior source.

Such simple steps resulted in the conclusion that the rotary engine is capable of achieving 20 000 hours of engine operation using current technology.

The most significant features of gas engines available now are their size and weight. Often based on existing ranges of diesel engines with high compression ratios, the available engines are almost invariably derated by a factor of 30 percent or more simply because of the volume of the fuel. The ongoing developments on gas injection systems have not yet come to fruition. Turbo-charging is the usual method used to diminish this problem to acceptable levels. The Wankel engine weighs about 60 percent of the equivalent piston engine and has a volume of about 50 percent. This, coupled with the ability of the engine to run at much higher speeds, provides an immediate solution to the main complaint of gas engine owners.

To fully explain both the advantages and limitations of the application of the Wankel engines to burn gas we need to consider the various aspects of design of the new generation of Wankel engine.

2.1 The LCR Family of Wankel Engines

The LCR family of engines is based on a modular concept in which the same carefully proportioned cross-sectioned module is used for both single and twin engines (3).

2.1.1 Materials

Wherever possible aluminium is used. The present design uses this material for all the main housings and stationary components. Steel is used for the liquid-cooled shaft and for the rotor. The trochoid housing is coated with Elnisil and the side-housings are coated with bronze. The durability offered by these coatings has been shown to be adequate for all normal applications.

2.1.2 Gas sealing

Gas sealing is undertaken with the most modern concepts available. The components are simple and durable and primarily made of widely available steels and cast irons.

2.1.3 Cooling

The trochoid housing and side-housings are cooled by water pumped by an integral pump fixed directly on the eccentric shaft. The cooling circuit provides a radial flow which allows maximum cooling from the areas of most heat stress allied with the minimum quantity of cooling fluid to reduce the working weight of the engine.

2.1.4 Rotor cooling

The cast steel rotor is cooled by contact with the water cooled side-housings, contact via the bearing with the liquid cooled shaft, and by the axial flow of charge air through the rotor, as shown in Fig 1.

2.1.5 Cooling of shaft and rotor bearing

Shaft and rotor bearing are cooled using the same fluid as is used for the side-housing and trochoid housing. Water is pumped through a tube down the centre of the hollow shaft by the same water pump as for the housings. Water flowing through the supply tube runs back down the shaft taking heat directly from the shaft at the most important points, which are the areas where combustion gases seep past the side seals, and close to the rotor bearing.

2.1.6 Lubrication

Lubrication is supplied using a total loss system with a metering pump. The consumption of lubricant is similar to all other internal combustion engines and the mechanism is highly reliable. The advantage of this system is that there is no sump, thereby reducing the weight of the engine and allowing the engine to work at any angle of inclination.

2.2 Potential

The result of these features is that the new range of Wankel rotary engines are extremely light-weight, simple in construction, cheap to manufacture and have the potential for exhibiting a durability of well over 20 000 hours and requiring no more maintenance than a conventional engine.

3. REQUIRED PERFORMANCE

The required performance of any engine will depend on the duty specification. Gas engines have in the past found niches which include:

(a) Large scale power generation.

(b) Portable and fixed irrigation equipment where electricity was either too expensive or not available.

(c) Small scale power generation where electricity was not available.

(d) Power generation in industries where both gas and electricity are available but tariff arrangements make the marginal cost of peak loads prohibitive.

(e) Heat pumps and co-generation systems.

(f) Power generation from biogas.

(g) Power generation from land fill gas.

This paper will address only the lower power ranges from 5 k.w. to 100 k.w. Within this range the main applications are heat pumps, co-generation systems, small scale power generation, and small irrigation and water supply pumps.

The durability requirements in this sort of application can mean a life of 20 000 hours with service intervals of 7 000 hours running time. The results of "South West Research Institute's" durability test show that there is no significant technical reason why the Wankel rotary engine should not provide such a lifetime and more. Indeed the Mazda 13B engine fitted in the Mazda RX7 car is one of the most reliable engines in its class.

The fundamental factor controlling the use of gas engines is the overall efficiency of the system. Present day gas piston engines are limited in compression ratio due to the problem of knocking and local over-heating caused by the faster flame propagation speed of the gas-air mixture when compared with petrol and diesel (4) and (5). The result is that the compression ratio is limited to below 10:1 whereas the rotary engine can use compression ratios of up to 12:1 without knocking, due mainly to the configuration of the combustion chamber. This allows efficiencies of in excess of 28-30 percent which is rarely equalled by existing gas burning piston engines below 100 k.w.

4. BASIC ADVANTAGES OF THE ROTARY ENGINE CONCEPT

4.1 Combustion chamber shape.

The Wankel rotary engine combustion chamber is in the shape of a sickle with a large length to width ratio, as shown in Fig 2. The result of this is that the rapid flame propagation experienced with gas produces in the Wankel rotary engine considerably less combustion noise along with reduced heat stress on the components, reduced peak mechanical stress on shafts and bearings and more complete combustion, than is experienced in piston engines.

4.2 Valves

The critical component in a piston gas engine when considering durability is the exhaust valve. The inlet valve will always be hot and the charge flowing into the engine is heated, resulting in reduced volumetric efficiency. The Wankel Rotary Engine inlet port is always cold as it is displaced from the combustion chamber and therefore the charge is not heated in the same way. The exhaust valve failure in gas piston engines occurs because of the lack of lubricating properties of the fuel. The fact that the Wankel Rotary Engine has no valves completely removes this problem. Furthermore, the rotary engine breathes more efficiently than a valved engine and the overall pressure drop through the engine is considerably reduced when compared with the equivalent piston engine.

4.3   Speed of rotation

Because of the basic configuration of the rotary engine it is possible to run at much higher speeds than a normal piston engine. This is due to the complete dynamic balancing of all the main components and also to the relationship between the rotor rotation and shaft rotation. It is normal to run a rotary gas engine at up to 5 500 r.p.m. whereas the normal limit for piston engines is 3000 r.p.m., or for larger engines 1 500 r.p.m.. This further increases the advantage in weight and size of the rotary engine in comparison with piston engines. The ease with which the rotary engine can cope with 2 pole speeds of 3000 r.p.m. or 3 600 r.p.m., is an important factor in matching the engine to the application.

4.4   Gas leakage

The possibility of crank case explosion caused by charge leakage past piston rings has always been an important factor in gas piston engine design, necessitating many special features. The rotary engines of the new generation have no sump in which such dangerous mixtures can collect and any charge gas seepage which passes the side seals enters the existing charge flow through the rotor and is fed directly back into the combustion chamber thereby requiring no other form of ventilation, as shown in Fig 2.

5.   EXISTING DESIGNS OF WANKEL ROTARY GAS ENGINE

5.1   The VMWP 225 Engine

Wankel R&D Gmbh developed this engine in association with Aisin Seiki during 1985 and 1986, as shown in Fig 3. The basic engine design is a 225 c.c. swept volume per revolution gas engine designed for a speed of 3000 r.p.m. producing a maximum power of 6.5 k.w., as shown in Fig 4. The engine has a charge cooled rotor and water cooled housings. The main housings are manufactured from cast iron as is the rotor and other main components apart from the shaft which is manufactured from steel. For the high durability required for such a design the housings were coated with hard chromium and the apex seals were manufactured from ceramics, as shown in Fig 5. The engine produced very good results, including efficiencies of over 30 percent at full-rated power and shows a potential durability in excess of 20 000 hours.

5.2   KKM 312 Engine

Wankel R&D Gmbh also developed in association with Gas Research Institute the KKM 312 c.c. Wankel Rotary Gas Engine with a rated power of 8.5 k.w. at 2600 r.p.m., as shown in Fig 6. The concept was originally for running in a co-generation system, as shown in Fig 7. The engine was again a cast iron design with charge cooled rotor and liquid cooled housing. The engine exhibited efficiencies up to 28 percent at rated power at 2500 r.p.m., as shown in Fig 8, Fig 9 and Fig 10, again showing the capacity for a service life in excess of 20 000 hours. The KKM 312 Engine was designed for all domestic and commercial applications such as pumps, generators and compressors.

5.3   Mazda 13B Engine

The durability of the Mazda 13B Gas Engine, as adapted by Wankel in association with South West Research and the Gas Research Institute, is in excess of 20 000 hours. This engine shows a remarkable performance providing 65 k.w. at 5 500 r.p.m. The maximum efficiencies are proven to be in excess of 29 percent with no significant modifications to the basic engine, as shown in Fig 11, Fig 12 and Fig 13.

5.4   The LCR Gas Engine

The new generation of Wankel Rotary Engines consisting of the LCR single and twin, share the basic advantages of the Wankel rotary concept with some significant advances, as shown in Fig 14. The liquid cooled shaft removes any problem related to heating of the shaft at part load and the likely durability of this engine should prove well in excess of 20 000 hours. The engine has been designed with light weight and low production cost in mind. The swept volume of the single LCR Engine is 407 c.c. and the twin 814 c.c., providing the potential for 18 k.w. for the single and 36 k.w. for the twin at 3,600 r.p.m. The design is suitable for co-generation applications. The weight of the single rotor engine is 29 k.g., and the twin 38 k.g.

The gas version is not yet developed though plans exist for such a development programme, to adapt the present range of petrol and diesel engines to burn gas.

6.   MARKETS

Markets for small gas engines have in the past proved elusive. The specific advantages of the rotary engine make it the most logical and elegant solution wherever an economic advantage for gas can be located. These areas will be found wherever electricity from a grid is not available but where gas is, which condition exists in some areas of the U.S.A. and New Zealand and elsewhere. Power generation, water pumping, air conditioning and other fixed applications are the primary areas of interest. Alternatively there are areas where heating and power can be provided by the supply of bottled gas or bio gas which will be a cleaner solution to power generation than diesel generators, with the rotary engine providing reduced maintenance costs and reduced initial equipment cost.

The other area where the gas rotary engine may come into its own is in countries where mains gas is available and electricity costs, particularly the peak cost, is very high and where the use of a co-generation system can provide very efficient production of useful energy for businesses and homes. The cleanliness of combustion of gas engines when compared with liquid fuelled machines, indicates a long-term future for gas engines. This allied to the likelihood of shortages of liquid fuels world-wide may result in the economic attraction of small gas engines improving in the future.

7. CONCLUSIONS

The Wankel rotary engine, when adapted to burn natural gas, exhibits all the usual advantages of the rotary engine concept and, in addition, the specific characteristics of the fuel when burnt in the normal rotary engine configuration, provides advantages in performance over and above that which can be expected from piston engines. These advantages accrue from the shape of the rotary combustion chamber which performs extremely well with natural gas; the absence of inlet and exhaust valves which is an important feature for durability; the maximum permissible speed of rotation; and the absence of a sump, a valuable safety factor.

The simplicity of design of the rotary engine and the likelihood of very high durability would indicate that for many applications the rotary engine is the correct technological choice. The rotary gas engine has yet to make its mark in large scale commercial production. Recent conferences have indicated a marked dissatisfaction with the available designs of gas engines. It remains to be seen if a manufacturer will take advantage of the availability of these designs.

The market for such an engine is susceptible to the fluctuating world economic situation. Any decision to go to production will depend on the various economic factors being in place.

The range of available designs is from 6.5 k.w. to 100 k.w. The designs all exhibit the attractive features of the Wankel rotary engine. The recent advances in production technology provide the possibility of manufacturing medium sized batches of engines at reasonably low cost which should compete on price effectively with the available piston engines.

There is every reason to suppose that the introduction of a range of Wankel rotary gas engines will prove successful both technologically and economically during the 1990s.

REFERENCES

(1) YAMAMOTO, K. Rotary Engine. Sankaido, Tokyo, 1981.

(2) KING, S. R. Durability of Natural Gas Filled Rotary Engines. S.A.E. International Congress and Exposition, Detroit, 1987.

(3) EIERMANN, NUBER and SOIMAR. The Introduction of a new Ultra-lite multi-purpose Wankel Engine. S.A.E. International Congress and Exposition, Detroit, Michigan, 1990.

(4) ABERNATHY, G. Natural Gas Engine adjustment and operation for maximum efficiency. Applied Engineering in Agriculture A.S.A.E., 1988, 4 (4).

(5) GILLESPIE, D.A. Gas-fed Reciprocating Engine Development. Proceedings of the Institution of Mechanical Engineers, 1987, 201 No. A2.

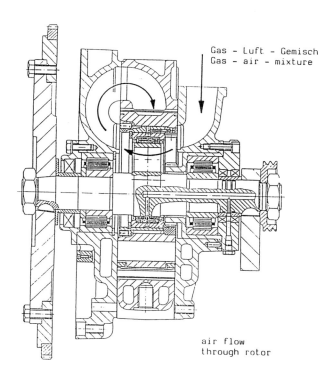

Gas - Luft - Gemisch
Gas - air - mixture

air flow
through rotor

Fig 1   Cooling using the charge-air method

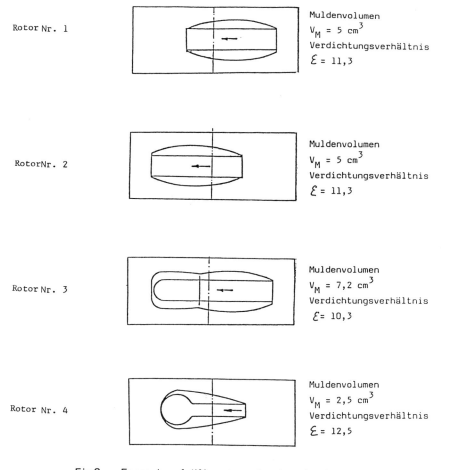

Rotor Nr. 1

Muldenvolumen
V_M = 5 cm^3
Verdichtungsverhältnis
$\mathcal{E}$ = 11,3

RotorNr. 2

Muldenvolumen
V_M = 5 cm^3
Verdichtungsverhältnis
$\mathcal{E}$ = 11,3

Rotor Nr. 3

Muldenvolumen
V_M = 7,2 cm^3
Verdichtungsverhältnis
$\mathcal{E}$ = 10,3

Rotor Nr. 4

Muldenvolumen
V_M = 2,5 cm^3
Verdichtungsverhältnis
$\mathcal{E}$ = 12,5

Fig 2   Examples of different combustion chamber
configurations

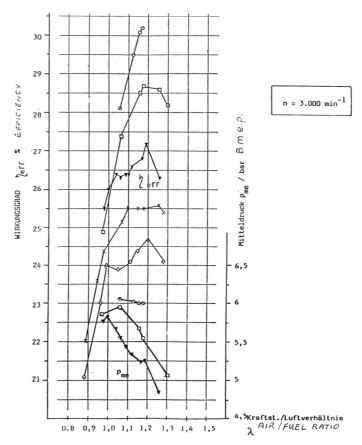

Fig 3   VMWP 225cc Wankel engine performance graph

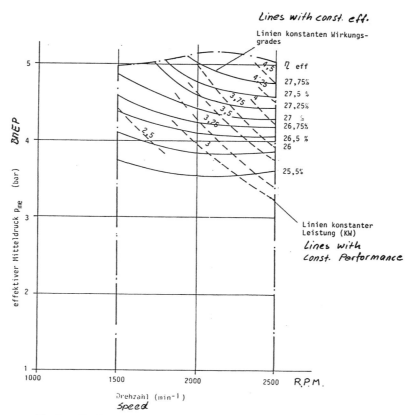

Fig 4   Performance curve for the VMWP 225cc gas engine

Fig 5   VMWP engine: assembly and components

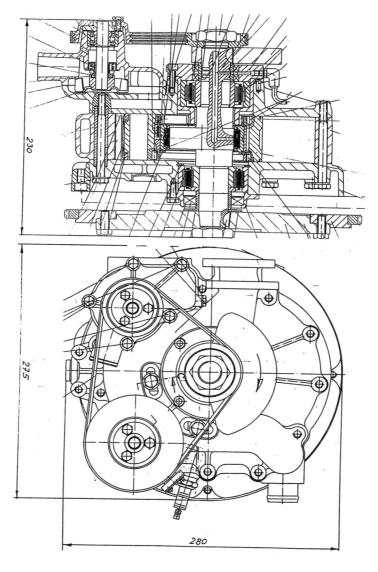

Fig 6   KKM 312 gas engine

Fig 7   Wankel gas-fuelled rotary engine: heat-pump unit

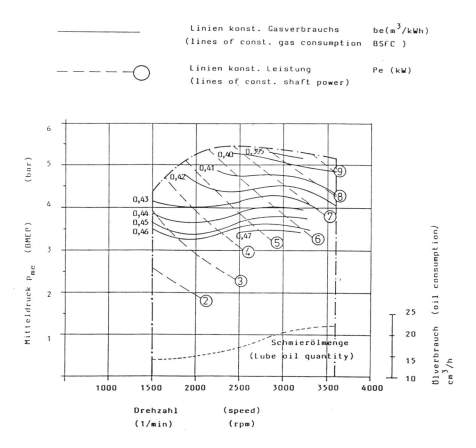

Fig 8   KKM 312 performance graphs

Vollast (W.O.T)

Stahllaeufer (steel rotor)
Verdichtung (compression ratio): 12

Drehzahl (speed)   2500 1/min
ZZP (ignition timing)   7 Grad v. OT   (BTDC)
Schmieroel (lube oil quant.)   30 - 35 ccm/h

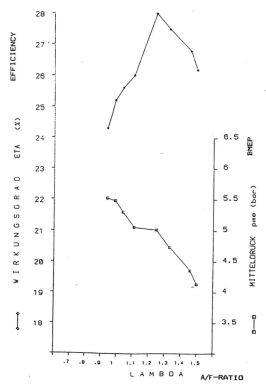

Fig 9   KKM 312 engine efficiency and bmep versus air/fuel
         ratio

Gasmischer
(Gas mixer)   CA 50

Zündzeitpunkt
(Ignition timing)   9,5° v. OT.

Verdichtung
(Compression ratio)   ε = 9,8

Umrechnungsfaktor
(Calculation factor) $\eta_e = \dfrac{10.035}{be}$ %; be (m³/kWh)   (BSFC)

η_e = efficiency

_____   Linien konst. Gasverbrauchs   be (m³/kWh)
                       (Lines of const. gas consumption BSFC)

----------O   Linien konst. Leistung   Pe (kW)
                       (Lines of const. shaft power)

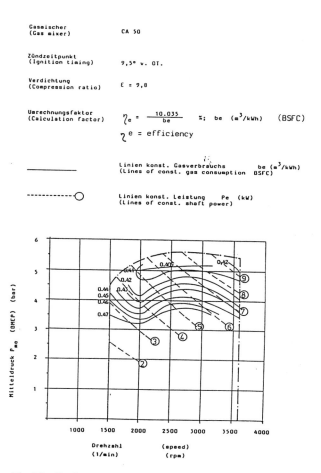

Fig 10   Performance graph of the KKM 312 gas engine

65

Fig 11   Mazda 13B engine, converted for gas, on the test
bench

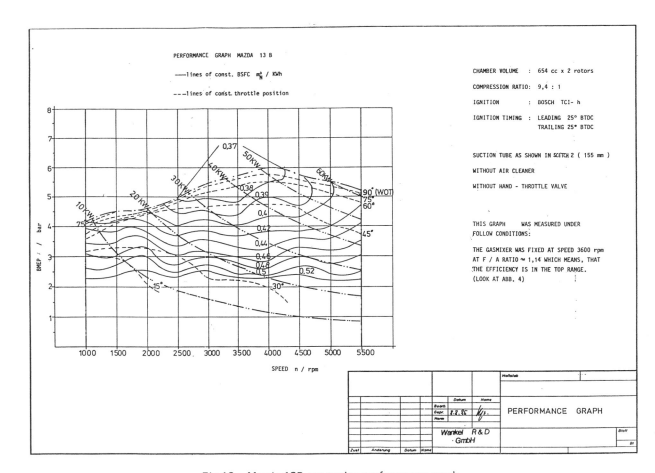

Fig 12   Mazda 13B gas engine performance graph

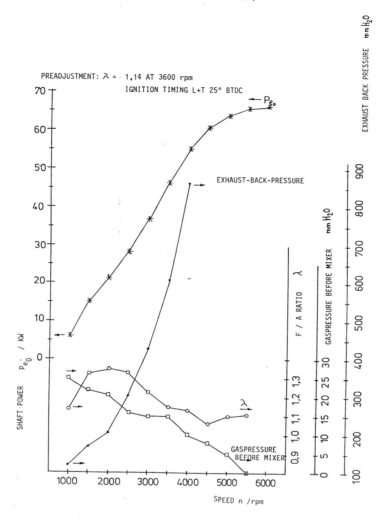

FULL - LOAD PERFORMACE GRAPH (WITH AIR-CLEANER)

Fig 13   Mazda 13B engine, converted for gas, on test bench

Fig 14   The LCR twin engine on which the gas derivative
is based

A model to predict performance and heat release of dual-fuel diesel engines

V PIROUZ-PANAH, PhD, CEng, MIMechE and **K AMIRASLANI**, MSc
Department of Mechanical Engineering, University of Tabriz, Tabriz, Iran

SYNOPSIS : In order to predict performance and heat release of dual-fuel diesel engines(D.F.D.E)a real engine cycle model was constructed and described here.In this model during whole cycle of events,heat transfer from the cylinder charge to the wall is considered and combustion process based on a synthetic single-zone combustion model with appropriate combustion pattern.The model predicts D.F.D.E performance and heat release rate with various proportions of gas in the dual fuel(diesel+gas).This model is of particular use as a research and development tool for designing of D.F.D.E and their conversion from pure diesel engine.

NOTATION

A	area
a	constant in equation(4)
act	constant in equation(3)
b	constant in equation(4)
c	constant in equation(4)and Annand equation coefficient
D	cylinder bore
DG	gas fraction in dual-fuel
E	internal energy
K	condutivity cofficient in equation (4) and constant in equation(2)
K'	constant in equation(3)
M	mass
m	constant in equation(2)
N	engine speed
P	pressure or rate of prepared fuel
Po_2	partial pressure of oxygen
Q	heat tranfered
q_{vs}	lower calorific of dual-fuel
R	rate of reacted fuel
Re	Reynolds number
T	temperature
V	volume
w	work
θ	crank angle

subscripts

a	air
d,f	dual-fuel
f	fuel
i	injected
g	gas
s	reference
u	unprepared
w	wall
t	total
1,2	Initial and end of step.

1 INTRODUCTION

Using gaseous fuels in intermittent interal-com-bustion(I-I-C)engines is not new concept . In this engines different types of gaseous fuels have been used successfully , and at present time small and large bore spark-Ignition engines(S-I-E)are opera-ting with gaseous fuels.However highly efficient compression-ignition engines(C-I-E) do not lend themselves for gas operation so easily due to their difference combustion characteristics.

In order to use gaseous fuels in C-I engines, different appraoches are used.These methods and their advantages and disadvantages are fully discussed in elsewhere(1,2)*.There it has been indicated that,most suitable method for using gaseous fuels in C-I engines is mixed diesel gas operation or so called dual-fuel diesel engine(D.F.D.E)with pilot injection.

In these engines during the suction stroke , a lean mixture of the gas and air is drawn through a suitable carburation system,into the cylinder , and near the end of the compresstion stroke , at injection time,a predetermined quantity of liquid diesel fuel,as a pilot fuel,injected into the hot gas air mixture and ignited it almost simultaneo-usly.It can be seen that with this method , engine design will not be altered , and its working cycle will be remained as highly efficient as dual cycle, or limited-pressure cycle,diesel engine.

2 LITERATURE REVIEW

Extensive literature survey in Reference(1)reveals that,research works on D.F.D.E are limited to some extend to experimental investigation(test bed type work),inorder to compare their performance with corresponding pure diesel engine(2-10).However it seems that on the development of mathematical models to predict D.F.D.E performance and heat release analysis least work has been done.However in recent years a few research workers constructed some special purpose models.

Karim and Zhaoda(11)constructed an analytical model for investigation of knock in D.F.D.E . Song et al(8)investigated D.F.D.E concept in a pre-chamber diesel engine with natural gas,both theo-reticaly and experimentaly.They studied combustion characteristics and heat release behaviour of pre-chamber diesel engines by employing a computer model.Again Karim et al(12)proposed an analytical model to examine the combustion of a fuel jet in a homogeneously premixed lean fuel air stream , in order to simulate the combustion in the D. F. D. Engines.To predict D.F.D.E performance a number of engine cycle models are also developed at TNO, Holland(13-14).These engine cycle models constructed mainly for their own engine development and opti-mization purposes.

*Numbers inside parentheses designates reference numbers.

The research presented in this paper is an extension of the work shown in reference (1).It forms part of a effort to predict performance and heat release rate of the real D.F.D.Engines. Reference(1)proposed an ideal engine cycle model for D.F.D.E.In that model heat transfer between cylinder charge and walls are not considered and combustion process considered to be instantaneous with unrealistic rates . In the present model , by introducing suitable combustion pattern and heat transfer correlation,deficiencies of the previous work were eliminated.

3 MODEL DESCRIPTION

As outlined before in D.F.D.E , during compression stroke a mixture of gas and air exists in the cylinder and at injection time a predetermined amount of liquid fuel (so called pilot injection) injected into the hot mixture of gas and air.After injection , liquid fuel jet breaks up into small liquid droplets and then they heated up to their boiling point and becomes as a vapour (physical delay)and then chemical reaction between fuel vapour and air starts (chemical delay) . Probably each droplet act as a ignition source and initiates local flame nuclei, which consumes nearby homogeneous combustible mixture . Since energy released from each droplet is considerable, so reliable and simultaneous ignition of the lean mixture of gas and air will be obtained.References(4 , 9)confirms this type of combustion phenomena in D.F.D.E.Also experimental investigation (10) showed that,depending on the amount of injected fuel,two types of combustion occur in these engines , one compression- ignition type and the other similar to that of the spark-ignition engines .When amount of injected fuel is high(or gas proportion is low) former type of combustion occurs,and with lower quantities of injected fuel latter forms of combustion exists.

After delay period , as a pure diesel engine, premixed combustion with higher combustion rate starts , since prepared fuel in D.F.D.E prior to combustion consists of gas plus injected fuel.This means that before combustion starts , greater part of the total fuel already exist within the cylinder. At the end of rapid combustion period,second stages of combustion starts,while rest of the liquid fuel still injecting.At this period, rate of combustion mostly is dictated by the oxygen availability. This type of combustion,called diffusion type combustion,and combustion rate during this period is lower than premixed combustion .After this period, while piston is moving towards B.D.C,expansion of combustion products occurs until exhaust valve opens.

As shown above closed cycle of D.F.D.E is similar to that of pure diesel engine cycle(15),but in D.F.D.E cycle,further complexity arises due to gas addition to the working fluid prior to combustion.

4 ASSUMPTIONS MADE FOR REAL CYCLE CONSTRUCTION

To construct real cycle of D.F.D.E following assumptions are made :
(a) Due to lean mixture of dual-fuel and air,dissociation of combustion products is not considered.
(b) Effect of residual gases from previous cycles is not considered.
(c) During combustion period , Single-Zone combustion model(15 ,16) modified and used.
(d) Ignition delay time taken as constant and equal to that of pure diesel engine,and it was not changed by variation of gas proportion.
(e) Internal energy and hence specific heat of

mixture calculated in terms of cylinder temperature and charge composition.
(f) With variation of gas proportion in dual-fuel, total input energy level per cycle will be remained constant and equal to that of the pure diesel engine.

5 DESCRIPTION OF EQUATIONS

In the proposed closed cycle,from IVC to EVO for each small time step,first law of thermodynamics written from reference(15)as :

$$dQ-dw = [E(T_2)-E_2(T_s)]-[E(T_1)-E_1(T_s)]-dM_f q_{vs} \qquad (1)$$

In this equation internal energy terms calculated in terms of temperature,which is based on reference temperature T_s,and composition of the cylinder change.Also calorific value of dual-fuel measured in this temperature.The $dM_f q_{vs}$ term is indicating amount of heat released from dual-fuel during each time step.

For calculation of dual-fuel mass burnt,dM_f , whitehouse and way (15 , 16)combustion pattern is modified and used as below :

$$P = K.M_{i.d.f}^{1-x} M_u^x . Po_2^m \qquad (2)$$

$$R = \frac{K'_{d.f}.Po_2}{N.\sqrt{T}} . \exp(\frac{-act}{T})\int(P-R)d\alpha \qquad (3)$$

Where $M_{i.d.f}$ includes mass of initial gas plus mass of injected fuel.Also $K'_{d.f}$ in equation(3) calculated as :

$$K'_{d.f} = CK' \qquad (4)$$

Where K' is constant coefficient for pure diesel case.Constant C is obtained from following polynomial function in terms of fraction of gas in dual-fuel as below :

for CNG $C = 0.9662 - 0.385GF+0.545GF^2-0.122GF^3$ (5)
for LPG $C = 1 + 0.0533GF + 0.717GF^2$ (6)

Coefficients of the equations(5)and(6)obtained by using Lagrange interpolation method.Due to different values of heating values of CNG and LPG,two seperate equations are used.

For calculating heat transfer term,Annand(17) expression is used :

$$\frac{dQ}{A} = a\frac{K}{D}(Re)^b(Tg - Tw) + C(Tg^4-Tw^4) \qquad (7)$$

The work term,calculated simply from mean cylinder pressure as :

$$dw = (\frac{P_1 + P_2}{2})(V_2 - V_1) \qquad (8)$$

6 NUMERICAL SOLUTION

Calculations start with known trapping condition (IVC)of cylinder charge(gas plus air).Specification of this point obtained from suitable stoichmetric calculations . For computational purposes whole closed cycle(from IVC upto EVO) is divided into small time steps and at each time interval with known initial condition of step , equation(1) solved for final condition of step using Newton - Raphson iteration method.Calculation procedure is similar to that of reference (18)except that in the present model during compresstion stroke , cylinder charge contains gas plus air composition , and whitehouse and way combustion pattern modified to take gas burning in to account.As stated above calculation ended when exhaust valve opens.At this stage performance parameters of D.F.D.E cycle such as,indicated power,Indicated thermal effeciency , indicated mean effective pressure , and indicated specific fuel consumption are calculated.

For the purpose of heat release analysis of the cycle,heat-release rate calculated from mass burning rate of dual-fuel($dM_f/d\alpha$)multiplied by heating value of dual-fuel.Mass burning rate of dual-fuel is calculated either from preparation rate equation(2)or reaction equation(3).

The computer programe has been written in fortran 77,and executed on P.C.IBM - AT machine , at university of Tabriz.

7 RESULTS AND DISCUSSION

For computational purposes,technical data of a typical D.I.diesel engine is used.Engine specification is given in Appendix(1).With this standard data, computation performed for pure diesel engine.Results obtained from this computation are used throughout this chapter for comparison purposes . It must be rememberd that in all cases , the level of total energy released per cycle from dual-fuel is constant and equal to that of pure diesel case.

Figure 1 shows the rate of heat released curves for D.F.D.E and pure diesel engines.As indicated the crank angles at which first and second peaks occur,are nearly same.But with increasing gas proportion,first peak increases and second peak becomes less distinctive.For example , for 30 percent LPG the maximum heat release rate is about 1.4 times higher than pure diesel engine.This increases to 2.4 in the case of 90 percent LPG substitution.As stated before with lower percentage of gas in dual-fuel,the combustion process follows the combustion characteristics of pure diesel engine, and combustion of mixture of dual-fuel and air is almost simultaneous.As percentage of substituted gas increases the combustion process tends to be similar to the S-I engines with probably flame propagation.This results was qualitatively confirmed by the results of Song(8).

Also from Figure 1 it can be seen that by increasing gas proportion , the rapidity of rise of the rate of heat release increases,which could be responsible for onset of"diesel knock"in D.F.D. Engines.In this respect less gas substitution is prefered.For this reason in this studies,for comparison purposes,only 30 perecent LPG is considered,since this is the limit which could be obtained in practice.(2)

Figure 2 compares cumulative mass injected and mass burnt curves for D.F.D.E(with 30 percent LPG) and pure diesel operation.It can be seen that initialy total injected mass of D.F.D.E is greater than pure diesel engine but after injection ceased, it becomes less .This is because when injection starts at 21 BTDC,30 percent mass of gas already exist in the cylinder and after injection , due to higher calorific value of dual-fuel,less mass of it will be required.In the case of total mass burnt curves initially due to higher burning rate of premixed type combustion,amount of total mass burnt is greater,but in later stages due to lower combustion rate,it becomes less than pure diesel case. However difference between cumulative mass injected and mass burnt in either case is nearly constant.

In Figure 3 cylinder pressure and temperature curves of D.F.D.E and pure diesel engines are compared.It can be seen that with 30 percent LPG in dual-fuel difference in both engines is not considerable,and peak pressure and temperature occurs at same crankangle respectively $8°$ and $35°$ ATDC. During ignition delay period,due to higher specific heat of cylinder charge,temperature and pressure of D.F.D.E is lower than pure diesel engine. Soon after combustion starts,due to fast burning of premixed charge of cylinder,rate of pressure and temperature rise in D.F.D.E is greater than pure diesel engine and hence peak temperature and pressure are also higher.During expansion stroke because of

lower rate of diffusion combustion , both temperature and pressure of D.F.D.E is lower than pure diesel engine.However these difference are magnitudely so small,that,they can be ignored.

Figure 4 compares indicator diagrams of D.F.-D.E and pure diesel engine at full load with similar injection timing.Trends in this figure is outlined in Figure 3.

Figure 5 compares indicated power output and indicated specific fuel consumption of both engines in consideration.It can be seen from Figure 5-a that,within useful range of engine speed D.-F.D.E produces slightly more power than pure diesel engine,but at very higher speeds it tends to become smaller.However difference is very small and less than 1 percent.Figure 5-b showes that isfc of D.F.D.E is always less than pure diesel engine. As staded in Figure 2 this is due to higher calorific value of dual-fuel,which results in using less mass of fuel.These results are qualitatively confirmed by reference(2).

It worth to note that,present model can predict performance of D.F.D.E for various gaseous fuels with different proportions . Figure 6 shows this effect on cylinder peak pressure and temperature for two types of gases.It can be seen that by increasing gas proportion in dual-fuel,peak pressure and temperature will be increased .The difference between two type of gases is due to their different calorific values.In practice sudden increase of peak pressure and temperature will increase mechanical and termal loads on the power train section.In this respect Figure 6 shows that , higher percentages of gas must be avoided.

8 CONCLUSIONS

On the basis of the previous discussions the following conclusions can be deduced :
(a) For using gaseous fuels in C-I engines , mixed diesel gas operation or D.F.D.E concept is most suitable method.
(b) By increasing gas proportion,maximum rate of heat release increases rapidly,which could be responsible for onset of"diesel knock",on D.-F.D.E,so less gas substitution is advised.
(c) With low percentage of substitute gas , power output of D.F.D.E is more or less same as pure diesel one,but gain in fuel economy is considerable.
(d) Due to fast burning of premixed charge,rate of combustion of D.F.D.E is higher,and this leads to greater cylinder peak pressure and temperatur.In practice this causes higher mechanical and thermal loads on cylinder assembly.In this respect higher percentages of substituted gas must be avoided.
(e) present model can predict performance and heat release of any types of D.F.D.E with various types of gaseous fuels.Also this model is of particular use as a research and development tool for designing of D.F.D.E and their conversion from pure diesel engine.

REFERENCES

(1) Pirouz-Panah,V.and Asadi,Y.Investigation of dual-fuel diesel engine with particular reference to engine cycle model.Journal of Engineering I.R.I,1989,5,44-49.
(2) Tiedema,P.,Van Der Weide,J.,Dekker,H.J. Converting diesel engines to the use of gaseous fuels.First International Gas Conference,Sri-Lanka,1982.
(3) Karim,G.A.and Burn ,K.S .The Combustion of

gaseous fuel in a dual-fuel engine of the compression ignition type with particular reference to cold intake temperature condition.
SAE paper 800263,1980.

(4) Moore,N.P.W.,Mitchel,R.W.S.Combustion in dual-
fuel engines.Joint conference on combustion,
ASME/I.Mech.E.1956,Ⅳ,300-307.

(5) Picken,D.J.,Harris,L.,Collins ,O.C.,Applica-
tion of biogas to commercial automotive vehi-
cles.EEC conference,Energy from Biomass,Ber-
lin,1982.

(6) Picken,D.J.,Few,P.C.,Smith ,R.J. The conversion of small diesel engines for use with biogas and natural gas.Leicester Polytenic internal report.

(7) Alimoradian,B.,Mockford,I.J.A dual fual system
for converting automotive diesel engines for
fuelling with natural gas.SAE.A Journal,1988,
39-45.

(8) Song,S.,Hill,P.G.Dual-fueling of a pre-chamber diesel engine with natural gas .Transac-
tions of the ASME,1985,Vol 107.

(9) Karim,G.A.A review of combustion processes in
the dual-fuel engine-the gas-diesel engine .
Prog Energy combust Science,1980,Vol 6,277-
285.

(10) Karim,G.A.,Raine,R.R.,Jones,W.An examination
of cyclic variations in a dual-fuel engine.
SAE paper 881661,1988.

(11) Karim,G.A.,Zhaoda,Y .An analytical model for
knock in dual-fuel engine of compression ignition type.SAE paper 880151,1988.

(12) Karim,G.A.,Kibrya,M.,Lapucha ,R.and Wierzba,
I.Examination of the combustion of a fuel Jet
in a Homogeneously premixed lean fuel-air
stream.SAE paper 881662,1988.

(13) Van Der Weide,J.,Seppen,J.J.Advanced hardware
and combustion technology for gaseous fuels
at TNO.Conference Gaseous Fuels for Transportation,Vancouver,1986.

(14) Seppen,J.J.Optimizing gas-fuelled engines
using modern computer techniques and dedicated fuel supply systems.International conference on New Developments in power Train and
Chassis Engineering,Strasbourg,1987.

(15) Benson,R.S.,Whitehouse,N.D.Internal combustion engines,1979.79-84(Pergamon Press).

(16) Whitehouse,N.D.,Way,R,J.B.A simple method for
the calculation of heat release in diesel engines based on fuel injection rate .SAE paper
710134,1971.

(17) Annand,W.J.D.Heat transfer in the cylinder of
reciprocating internal combustion engines.
Proc.I.Mech.E.1963,177,973.

(18) Pirouz-Panah,V.,Jafari,M.,Investigation of
combustion models in D.I diesel engines with
particular reference to single-zone model.N.
I.O.C conference on oil,gas,and petrochemis-
try,Tehran,1989(N.I.O.C)

APPENDIX 1

Engine specification
Engine type 4-stroke,D.I.diesel engine ,Naturaly
Aspirated

Bore	0.115 m
Stroke	0.140 m
Displacement volume	8.72 lit
Compression ratio	16.8
Rated power at 2300rpm	135 kw
Ignition delay period	1-2 ms
Injection timing	21 BTDC
EVO	123 ATDC
IVC	118 BTDC
cylinder number	6

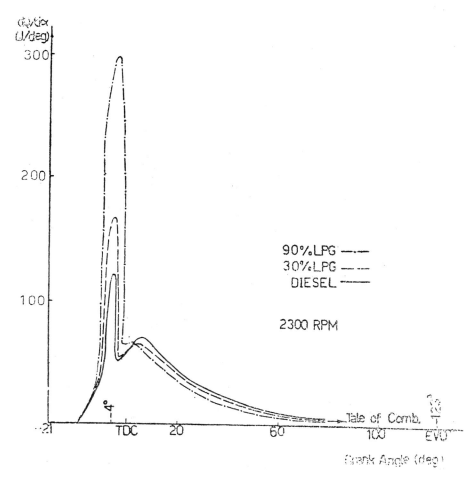

Fig 1   Comparison of heat release rate for pure diesel
operation and dual-fuel with two LPG percentages

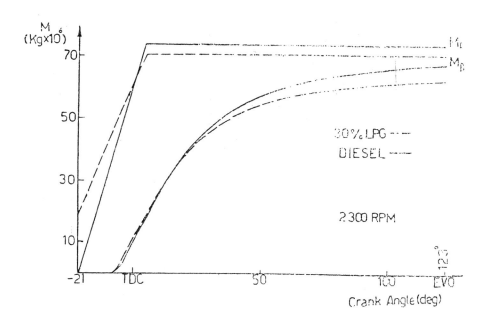

Fig 2   Comparison of cumulative mass injected and mass
burnt for dual-fuel and pure diesel operation

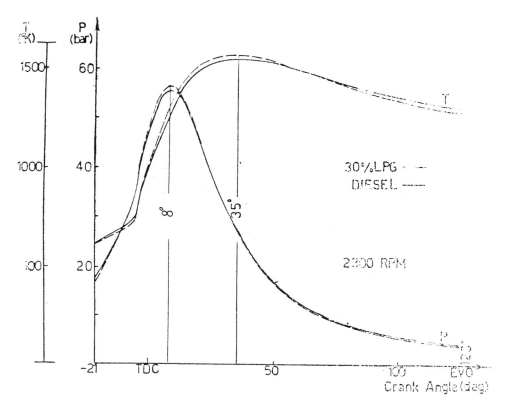

Fig 3   Comparison of cylinder pressure and temperature
for dual-fuel and pure diesel operation

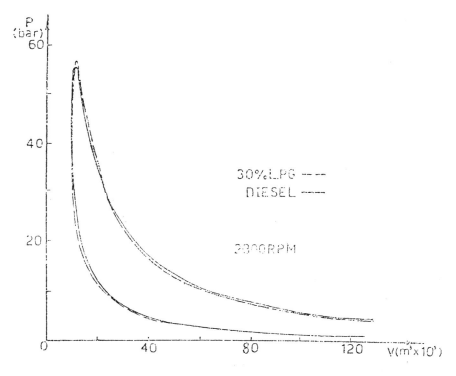

Fig 4   Comparison of P—V diagrams for dual-fuel and
pure diesel operation

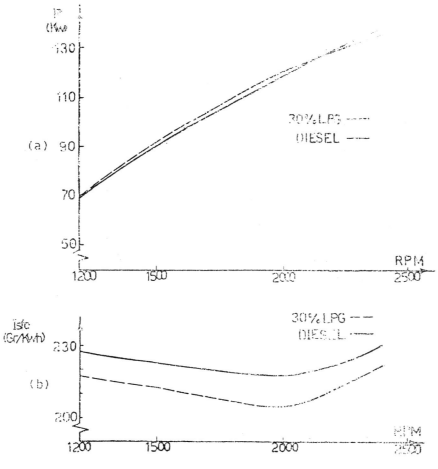

Fig 5   Comparison of indicated power output and isfc
         curves for dual-fuel and pure diesel operation

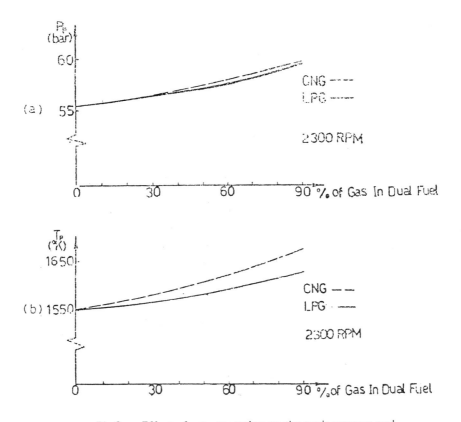

Fig 6   Effect of gas proportion on the peak pressure and
         peak temperature of dual-fuel diesel engine

An investigation of mixing and combustion in a lean burn natural gas engine

S J CHARLTON, D J JAGER, M WILSON and **A SHOOSHTARIAN**
School of Mechanical Engineering, University of Bath, UK

SYNOPSIS   Initial results from an experimental study of mixing and combustion in a lean burn natural gas engine are presented. The findings of this study confirm that dramatic reductions of NOx are possible by this method. NOx emissions have been reduced from 15-20g/kWh to 0.5-2.0g/kWh at the leanest air-fuel ratios. The paper includes a parametric study of NOx emission with the quality and quantity of mixture supplied to the prechamber. It is shown that the prechamber is itself a significant source of NOx when the main chamber is operated lean. The emission of NOx with variation of ambient temperature over the range 5-40degC is presented.

1   NOTATION

C_p	specific heat at constant pressure	J/kgK
m	mass flow rate	kg/s
p,P	pressure	N/m^2
P_{pc}	prechamber supply pressure	N/m^2
P_{man}	intake manifold pressure	N/m^2
R_{pc}	compressor pressure ratio	
R_{pt}	turbine pressure ratio	
T	temperature	K
T_c	intercooler coolant temperature	K
V_a	volume of prechamber air/cycle	m^3
V_{pc}	volume of the prechamber	m^3
ε	intercooler effectiveness	
γ	ratio of specific heats	
η_c	compressor efficiency	
η_t	turbine efficiency	
κ	$(\gamma-1)/\gamma$	
λ	prechamber filling ratio	

2   INTRODUCTION

The paper describes initial results from an experimental research programme on the mixing and combustion processes of a divided chamber, stratified charge natural gas engine. The experimental work is being carried out on a single cylinder version of the Dorman SE multi-cylinder diesel and natural gas engine, which has a swept volume of 3.82 litres/cylinder and a maximum operating speed of 1800rev/min. The single cylinder research engine was designed at Bath University(1)[1] where it is installed. The engine is shown in figure 1.

The combustion system consists of two chambers connected by a number of small nozzles. The smaller chamber, which accounts for only 2-3% of the clearance volume, is located centrally in the cylinder head.

The aim is to produce a mixture close to the stoichiometric air-fuel ratio in the smaller prechamber and a lean mixture in the main chamber. Ignition occurs by an electrical spark in the prechamber, the resulting flame spreading into the main chamber to ignite the lean mixture. Figure 2 shows the prechamber, the

spark plug and the direct air-gas supply through which mixture is admitted via a self-acting disc valve. The excess air in the main chamber acts as a diluent and reduces the resulting combustion temperature, thereby making possible a reduction in the level of NOx produced.

The prechamber content will be readily ignited, being within the spark flammability limits, and will provide a dispersed high energy

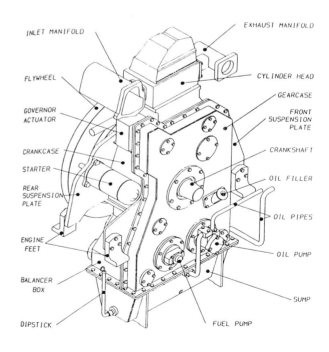

Fig 1   Diesel version of the single cylinder research engine

---

[1] References appear at the end of the text

source for the lean mixture in the main chamber. This approach should improve cyclic dispersion and extend the lean misfire limit in the main chamber by virtue of the consistently high ignition energy.

Natural gas is a mixture of permanent gases, the composition of which depends on the source. The main combustible component is methane (CH_4) but other combustible gases such as ethane (C_2H_6), propane (C_3H_8) and butane (C_4H_{10}) may be present in small, but significant, quantities. Some sources produce natural gas having a high non-combustible content, usually CO_2 and N_2. In this study the fuel was drawn from the UK mains natural gas supply which, subject to small variations in composition and calorific value, has the properties listed in Appendix A.

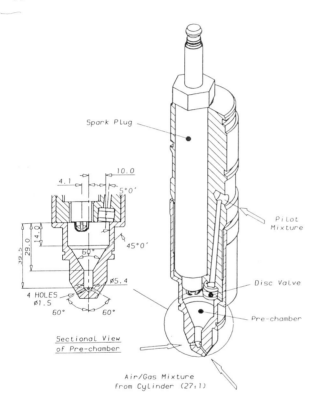

Fig 2   Prechamber detail and perspective section view

Since the early 1980s there have been a number of research programmes, mainly in the USA and Europe, on the development of low NOx combustion systems for natural gas engines, in response to demands for cleaner engines. For example, the application of exhaust catalysts (2), exhaust gas recirculation and special spark plugs, such as the 'Catplug' and 'Swirl Chamber' plug (3). Other workers have studied aspects of the stratified charge approach, for example Ref.(4), who investigated the effects of prechamber volume, air-fuel ratio and nozzle configuration on the performance and exhaust emissions of a gasoline engine. Possibly the earliest publication on this subject was by Serve (5) of Superior Engines who in 1982 outlined a stratified charge design for reduced NOx. Snyder *et al* (6) have recently published

results from a photographic study of stratified charge combustion in a simulated natural gas engine. Waukesha (7) have published an account of the development of the Sulzer AT25, for operation with a stratified charge combustion system for natural gas operation.

3   COMPUTATIONAL STUDY OF MIXING IN THE PRECHAMBER

The computational fluid dynamics code PHOENICS has been used to investigate the nature of the processes in the prechamber prior to ignition. The development of the model was described in an earlier paper (8) and will not be discussed in detail here. The model represents the prechamber only, with a time-varying pressure boundary condition applied at the gas valve and at the nozzles which connect the prechamber to the cylinder. The cylinder boundary condition was obtained from a thermodynamic model of the whole system. The simulation starts at inlet valve opening (ivo) with the assumption that the prechamber contains only residuals from the previous cycle at a temperature of 800K, burned at the stoichiometric air-fuel ratio. The solution is allowed to run from ivo through to firing tdc, with no attempt to model combustion.

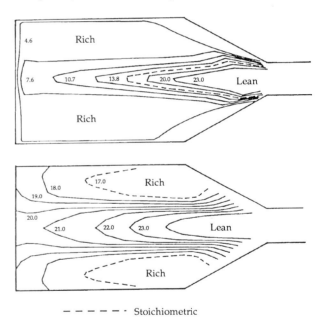

– – – – – · Stoichiometric

Fig 3   Spatial variation of air—fuel ratio in the prechamber at 12 degrees btdc predicted by the PHOENICS CFD code. *Upper*—with a pilot feed of pure gas. *Lower*—with a 9:1 air—gas pilot feed

Figure 3 shows the spatial variation of mixture strength in the prechamber at 12deg btdc for both a pure gas pilot feed and a 9:1 air-gas pilot feed. With the pure gas feed, figure 3a, the air-fuel mixture which develops around the outer third of the prechamber is very rich, varying between 4:1 and 8:1. In the central third of the prechamber the influence of the incoming lean jet has had a small effect, to the extent that a zone which is leaner than stoichiometric exists in the lower part of the chamber. The distribution of the mixture also depends upon the pilot supply pressure and the quantity of pilot mixture supplied. In this case the pilot supply pressure was quite high compared with the intake manifold pressure, which sets the cylinder pressure during induction, and a high degree of residual scavenging occurred. With a lower pilot supply pressure the distribution will be altered slightly by the greater residuals content.

When a pilot charge having an air-fuel ratio of 9:1 is supplied, the distribution of mixture strength is quite different, as may be seen in figure 3b. The chamber contains a spread of air-fuel ratio from 9:1 to 24:1, with the greater part of the content in the range 17:1 to 24:1.

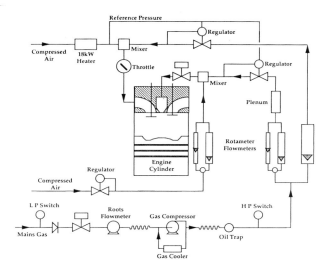

Fig 4   Gas and air supply and control systems for the single cylinder research engine

4   EXPERIMENTAL   STUDY

Figure 4 shows a schematic representation of the complex gas and air supply arrangements. With this configuration the air-fuel ratio to either chamber may be varied continuously over the range of interest by fine adjustments of flow control valves located outside of the test cell. Engine speed control was accomplished by a Heinzmann electronic governor which operates on the throttle placed downstream of the main gas mixer. Gas flows are measured by 'Rotameter' variable area flow meters, calibrated for natural gas at a reference pressure and temperature. The actual pressure and temperature at each flow meter are measured and used to correct the readings. Prechamber gas and air flows may be varied over a wide range. To cope with this, without sacrificing accuracy, two flow meters are used to cover the flow range. Total gas flow is measured by a positive displacement 'Roots' meter which has an accuracy of ±1% over the working range. Air flow is measured using a flat plate orifice installed and corrected in line with BS 1042 (9). Gas and air pressures are measured by water or mercury manometers and Bourdon tube pressure gauges. Temperatures are measured by chromel-alumel thermocouples. Oxides of nitrogen are measured by a chemiluminescent analyser with bottled air and span gases for calibration.

High speed in-cylinder data is recorded by an A/D card and software mounted on a IBM compatible micro computer. The card is capable of sampling a single channel at up to 1MHz, or it may be multiplexed to read up to 16 channels at a reduced rate. This system is capable of recording up to 150 cycles of data at an engine speed of 1500rev/min.

All of the experimental data taken during the study were obtained with the engine operating under steady-state load and speed conditions. The results presented in this paper were obtained with the following parameters held constant.

Table   1   Engine   Parameters   for   the Experimental   Programme

Engine Speed	1500rev/min
Turbocharger Efficiency	55% (simulated)
Intercooler Effectiveness	70% (simulated)
Ambient Temperature	25degC (simulated)
Compression Ratio	9.77:1
Prechamber Volume	10cm3
	(2.3%of clearance volume)
Swept Volume	3820cm3
Bore	160mm
Stroke	190mm

5   SIMULATED   TURBOCHARGING

In order to simulate the turbocharger and intercooler of the multi-cylinder engine, it is necessary to assume values for the efficiencies of the individual components, and for the ambient pressure and temperature. The values chosen are typical of those achieved by the multi-cylinder engine, and may be found in Table 1. During a test, it is necessary before taking a full set of data, to establish that the simulated turbocharging system is operating in the required way. A schedule of boost pressure versus bmep, shown in figure 5, is used to establish both the boost pressure and the boost temperature. The latter is calculated from the assumed ambient temperature, compressor efficiency, intercooler effectiveness and intercoller coolant temperature, as shown in figure 6.

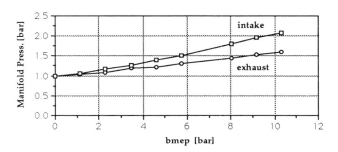

Fig 5   Schedule of boost pressure used for the simulation of turbocharging. Typical exhaust pressure shown for reference

An electrical heater installed in the intake pipe work is used to preheat the air to the temperature for boost simulation. The heater is controlled by a proprietary three term controller in a closed loop. When the intake side is set up, the exhaust back pressure is set to an initial value, shown in figure 5, and the quasi-steady turbocharger efficiency is calculated from a set of measurements, which include the 'turbine' entry temperature in the exhaust manifold. Small adjustments are then made to a variable restriction placed in the exhaust pipe work in order to bring the simulated turbocharger into power balance. The detailed design of the simulated turbine is reported in Ref.(1). It consists of a fixed orifice, a plenum and a variable flow restriction in series, as defined in Ref.(10).

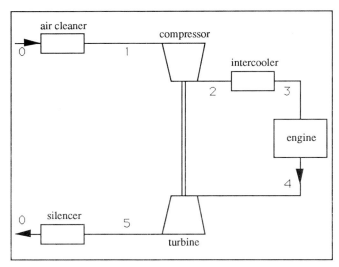

$$T_3 = T_0 \left[(1-\varepsilon)(1 + \frac{(R^{\kappa_1} - 1)}{\eta_c}) \right] + \varepsilon T_c$$

$$\frac{p_4}{p_5} = R_{pt} = \left[\frac{(mC_p T)_1 \, \eta_t \, \eta_c}{(mC_p T)_4 \, \eta_t \, \eta_c - (mC_p T)_1 \, (R^{\kappa_1} - 1)} \right]^{\frac{1}{\kappa_4}}$$

Fig 6   Relationships used in the simulation of turbocharging to determine the required manifold temperature and exhaust back pressure

$$\lambda = \frac{Volume\ of\ prechamber\ charge\ at\ manifold\ pressure}{Volume\ of\ prechamber}$$

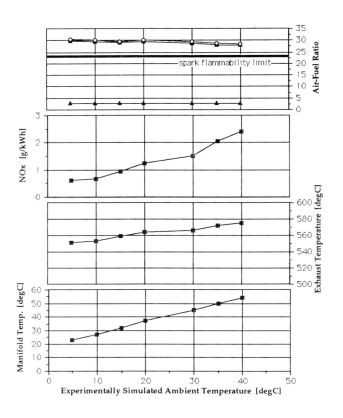

Fig 7   Response of the engine to variation of ambient temperature

6   DISCUSSION OF RESULTS

Figure 7 shows the response of the engine to variation of simulated ambient temperature, over the range 5 - 40 degC. This was achieved using the simulated turbocharging system as described in the previous section. The turbocharger efficiency and boost pressure were held constant during the test, and as a result, it can be seen that the main chamber air-fuel ratio decreased slightly with increasing ambient temperature. This effect combined with the increasing charge temperature combine to increase the emission of NOx from 0.6 to 2.3 g/kWh. Despite the increasing charge temperature, both the audible knock and thermal efficiency remained almost unchanged. The knock that could be detected was very light and since it did not appear on the main chamber pressure trace, was believed to be occurring in the prechamber.

Figures 8a and 8b show a map which is useful when considering the influence of the prechamber parameters on the performance of the engine and the production of NOx. The axes relate to the quantity of pilot charge (abscissa) and the quality of pilot charge (ordinate). Quantity is represented by a parameter known as the filling ratio which is defined as follows:

Thus a value of unity would be expected to just fill the prechamber if it completely displaced the residuals. Quality of pilot charge is represented by the supply air-fuel ratio. This should not be confused with the prechamber air-fuel ratio at the point of ignition since dilution from the main chamber will occur during the compression stroke. The diagram shows an operating envelope bounded by three constraints. At the lower edge the pilot supply is too rich to produce a flammable mixture in the prechamber at the point of ignition. At the upper edge the pilot supply is too lean, leading to misfire, instability and eventually stalling. The right hand edge of the envelope is limited only by the capacity of the pilot supply valve and the supply pressure relative to the manifold pressure.

Figure 8a shows contours of prechamber supply pressure normalised by manifold pressure (Ppc/Pman). Throughout much of the induction stroke, the pressure in the prechamber is below that in the manifold, since at pressure ratios below unity a filling ratio of 0.2 - 0.6 may be achieved. To achieve a filling ratio in excess of unity, a pressure ratio of 1.13 to 1.15 is required. Also shown in figure 8a are lines of constant prechamber air volume flow rate (Va/Vpc). The map was obtained by fixing the air flow rate and varying the quantity of fuel induced into the prechamber. Figure 8b is typical of the response of NOx emission from the engine with variation in prechamber charge quality and quantity. When operated towards the rich limit the NOx emission is at a maximum value of 4 - 4.5g/kWh. When run close to the lean limit the NOx emission is as low as 2 - 2.5 g/kWh. It is believed that the variation of NOx emission across the prechamber operating envelope is entirely due to NOx production within the prechamber itself.

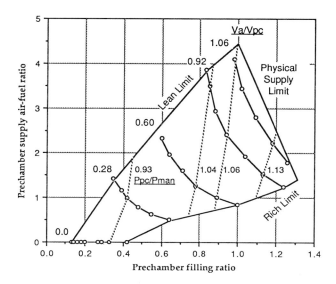

Fig 8(a) Prechamber performance map showing the feasible operating region and contours of supply pressure ratio and prechamber air volume flow rate.
bmep = 10.33 bar
Ignition 14 degrees early
Main chamber air—fuel ratio = 27.5:1

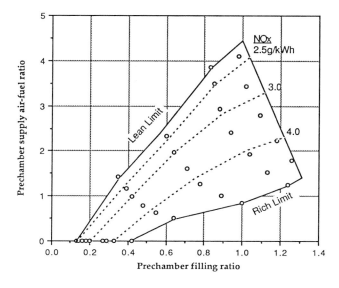

Fig 8(b) Prechamber performance map showing the feasible operating region and contours of NO_x production
bmep = 10.33 bar
Ignition 14 degrees early
Main chamber air—fuel ratio = 27.5:1

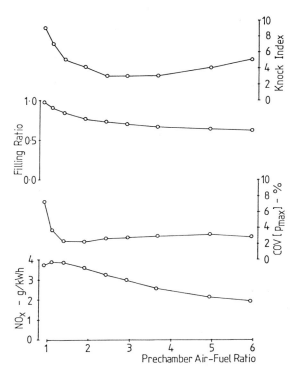

Fig 9 Variation of combustion related parameters with prechamber supply air—fuel ratio

Figure 9 shows the variation of prechamber knock, cyclic dispersion and NOx with prechamber air-fuel ratio. Knock index is an arbitrary quantity based on the subjective level of audible knock. The coefficient of variance (COV) is based on maximum firing pressure in the main chamber for a sample of 110 cycles. This figure underlines the disadvantages of operating the prechamber on a pure gas charge - the tendency to knock is increased, as is the roughness as measured by COV. NOx emission is seen to decrease in line with the map presented in figure 8b.

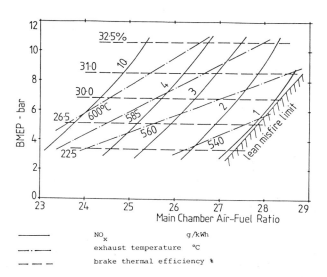

Fig 10 Main chamber performance map showing contours of brake thermal efficiency, exhaust temperature and NO_x. Prechamber conditions were held constant

Figure 10 presents a map of engine bmep against main chamber air-fuel ratio for an ignition timing of 12deg btdc and a prechamber air-fuel ratio and filling ratio of 3:1 and 1.00 respectively. Shown as contours on the map are NOx, exhaust temperature and brake thermal efficiency. Perhaps surprisingly the efficiency appears to be independent of the main chamber air-fuel ratio, it increases with load reaching a maximum of 32.5% at the highest load. At constant air-fuel ratio the NOx emissions increase with load, as intake temperatures rise due to increasing turbocharge. At the highest load with an air-fuel ratio of 28:1, which is just stable, NOx emissions of 2g/kWh are obtained. Given the 'diagonal' NOx characteristics on the load / air-fuel ratio plane, it would be desirable to design the fuelling system to follow such a contour – thereby allowing the engine a stable and drivable air-fuel ratio at part loads where good load acceptance will be essential.

7   CONCLUSIONS

The computational fluid dynamics code PHOENICS has been used to investigate the mixing processes in a divided chamber natural gas engine. Results from the computational study show that the mixture in the prechamber at the point of ignition, is extremely inhomogeneous which may contribute to cyclic dispersion.

Experimental results demonstrate that extremely low (<2g/kWh) emissions of NOx are possible with this system, by adopting main chamber air-fuel ratios of 27 to 29:1.

From the results of the study it appears that the best control strategy is to operate the prechamber with a pilot supply of both air and gas, in a ratio which depends on the main chamber air-fuel ratio, and with a pressure just high enough, in relation to manifold pressure, to almost completely scavenge the prechamber of residuals. A greater pressure would probably lead to pilot flow during the exhaust stroke and a possible loss of fuel to the exhaust valve.

When the main chamber is optimised for low NOx emission, the prechamber can become an important source of NOx, which may be influenced greatly by the choice of pilot air-fuel ratio.

Ambient temperature has a significant effect on the emission of NOx. When the ambient temperature is artificially varied between 5degC and 40degC the level of NOx is increased from 0.6g/kWh to 2.3g/kWh.

ACKNOWLEDGEMENTS

The principal author wishes to acknowledge the support of Dorman Diesels Limited of Stafford, England and The Science and Engineering Research Council for their support of this research project.

REFERENCES

1. **Moore D S,** ' Design of a Single Cylinder Research Engine and Development of a Computer Model for Lean Burn Combustion Studies',PhD Thesis, University of Bath, 1987.

2. **Klimstra J,** 'Catalytic Converters for Natural Gas Fuelled Engines – A Measurement and Control Problem', SAE Paper 872165, Toronto 1987.

3. **Klimstra J and Lukey C,** 'Improved Ignition System for Increased Reliability and Lower NOx Emissions of Gas Engines', 17th World Gas Conference, Washington DC, June 5-9,1988.

4. **Ryu, Chtsu and Asanuma,** 'Effect of Torch Jet Direction on Combustion and Performance of a Prechamber Spark Ignition Engine', SAE 870167, Detroit 1987.

5. **Serve J V,** 'NOx Reduction on Large Bore Turbocharged SI Engines',ASME 82-DGP-16, Energy Sources Technology Conference, March 1982.

6. **Snyder, Wright and Dexter S,** 'A Natural Gas Engine Combustion Rig with High Speed Photography', ASME Conference 1987.

7. **Pohl J M,** 'Design and Development of the Waukesha AT25GL Series Gas Engine', ASME Energy Sources Technology Conference, Jan 1988.

8. **Charlton S J, Jager D J, Wilson M and Shooshtarian A,** 'Computer Modelling and Experimental Investigation of a Lean Burn Natural Gas Engine', SAE Paper 900228, Detroit 1990.

9. **British Standard 1042,** Air Flow Measurement, British Standards Institute.

10. **McKenzie E M J and Dexter S G,** 'The Use of a Single Cylinder Test Engine for Research and Development of Meduim Speed Diesels', Ricardo Unrestricted Report DP 82/1667,1982.

11. **Parry R D,** Communication with British Gas, October 1988.

APPENDIX   A
Properties of UK Mains Natural Gas (11)

Component		%Volume	
N2	Nitrogen	1.17	
CO2	Carbon Dioxide		0.37
CH4	Methane	94.40	
C2H6	Ethane	3.50	
C3H8	Propane	0.40	
iso-C4	iso-Butane	0.04	
n-C4	n-Butane	0.06	
C5	Pentanes	0.03	
Octane rating		≈120	
Calorific Value		38.7MJ/m3	
Stoichiometric Air-Fuel Ratio		16.5 kg/kg	

A natural gas fired two-stroke engine designed for high thermal efficiency and low environmental impact

J R GOULBURN, MIMechE, **G P BLAIR**, FIMechE, FSAE, FEng, **M DONOHOE**, MSc, GIMechE
Department of Mechanical and Manufacturing Engineering, The Queen's University of Belfast, UK

SYNOPSIS

A two-stroke cycle crankcase compression engine offers mechanical simplicity, low cost and favourable power/weight ratio. A natural gas fuelled, stratified charge, two stroke engine has been developed for efficiency, waste heat recovery and minimum environmental impact. The development programme and the engine performance is described in some detail, together with a discussion of the engine's continuing development.

1 INTRODUCTION

Although the engine described in this paper was not in the first instance designed for co-generation applications, its characteristics make it an interesting candidate, as can be seen from a summary of the original duty. This was as a natural gas fuelled compressor drive for a heat pump, with waste heat from the engine augmenting the heat pump condenser output.

A small (2.5 kW) engine was specified. Ideally it was to be low-cost, compact and quiet enough for internal domestic applications. Engine thermal efficiency was to be high, with a target figure of 30% (lower heating value or net calorific value basis). It had to be capable of continuous running, with maintenance once per year. The engine cylinder and exhaust system were to be water cooled for waste heat recovery purposes, and exhaust gas constituents had to be environmentally acceptable.

Many of the above requirements can be achieved by converting an existing four-stroke engine with a high compression ratio to natural gas fuel, and this is the conventional solution. There are some disadvantages however, such engines are relatively bulky, expensive and mechanically complex. Problems can occur with poppet valve seats, particularly with continuous running. The converted four-stroke engine approach had been successfully demonstrated (1) and it was felt that a two-stroke engine offered the possibility of a better solution. Consequently, a two-stroke engine was designed and built, in an attempt to meet the above specification.

The basic two-stroke cycle yields one power stroke per revolution, compared to every two revolutions for a four-stroke. This gives an immediate potential advantage of two on the power/weight ratio for a two-stroke. Piston porting is used, consequently poppet valve operating mechanisms are not required, and it also follows that poppet valve seat wear cannot occur.

On the other side of the coin, two-stroke engines have the reputation of being noisy, inefficient, difficult to lubricate and with dirty exhaust gas. There have been a number of developments in recent years designed to deal with these problems, and these advances have been incorporated where possible in the engine described here.

2 DESIGN PHILOSOPHY

A crank case compression two-stroke engine with piston porting has traditionally been used as a cheap and compact source of power. Its principal disadvantage has been the nature of the scavenging process.

In this process, for much of the time that the exhaust port is open to discharge the combustion products, the inlet ports which are transferring the fresh charge to the cylinder are also open. A significant portion of the fresh charge may be lost by a short circuiting effect through the exhaust port.

This leads to a reduction in thermal efficiency, and high hydrocarbon levels in the exhaust gases.

Various techniques have been developed to prevent, or minimise, the loss of fresh charge, and in recent years, two-stroke engines have been developed with thermal efficiencies and exhaust gas hydrocarbon levels comparable to four-strokes (2). The developments have been directed towards gasoline fuelled engines, but this does not prevent their application with a gaseous fuel. Indeed, some of the problems presented by a liquid fuel which must be evaporated before being completely mixed with air, do not arise with natural gas.

The most profitable areas of development have been (a) novel deflector piston shapes and transfer port arrangements (b) stratification of air and fuel (c) fuel injection after or very shortly before exhaust port closure.

In the design described here, (a) and (b) were used in the first instance. The decision to use the (b) approach rather than (c) was because of its low cost and mechanical simplicity, which was a part of the specification.

Figure 1 shows the shape of the deflector piston, which had been developed earlier by Blair (3). It can also be used to illustrate the detail of how fuel and air stratification is achieved.

Natural gas enters the engine through a tapered needle throttle valve, and a non-return reed valve. When the piston rises, a subatmospheric pressure develops in the crankcase, which then induces air through a piston operated port. The flow of air can be controlled by means of a throttle in the air intake system.

When crankcase pressure is subatmospheric, natural gas is induced into the back transfer duct. The duct volume is such that the natural gas charge does not reach the crankcase. When the piston descends, it raises the crankcase pressure, and at the appropriate positions opens the air ports and then the gas port.

The air streams and the gas stream are initially completely separated, and mixing takes place in the cylinder. The shape of the deflector piston promotes rapid mixing by a 'squish' action near TDC. When the air ports open, there will be a significant loss of air through the exhaust port, but the forcing pressure difference between the crankcase and the cylinder will decrease, and the idea is that when the gas port opens, the gas will enter the cylinder slowly. It was accepted from the outset that the gas port timing would need to be optimised by experiment, for this reason the gas port timing could be altered by means of inserts (Figure 1).

3 DESIGN DETAILS

The main features of the design are illustrated in Figures 1 and 2. The engine cylinder is water cooled, and the cylinder head designed to enable different trapped compression ratios. The crank is overhung, thus enabling both main bearings to be sited in a single oil bath, whilst the piston is a deflector type using the design developed by Blair (3). As the engine is a prototype, it has been machined from solid aluminium.

Engine geometrical details are:

Bore	5.7 mm
Stroke	5.7 mm
Swept volume	145.45 cm^3
Compression ratio	10.45:1 (trapped)
	12.36:1 (geometric)
Exhaust port opens	125° ATDC
Air transfer port open	141° ATDC
Air inlet port opens	48° BTDC
Ignition	fully electronic
Scavenging	stratified charge, hybrid

standard domestic gas meter. A solenoid cut-off valve was included in the gas supply line, which only opened when the ignition system and dynamometer were switched on. Figure 1 shows how the gas enters the inlet manifold, via an adjustable tapered needle throttle and a one way reed valve to prevent back flow of the gas during crankcase compression.

Airflow rate is measured by means of a standard orifice plate attached to an airbox, in conjunction with a micromanometer.

Engine lubrication was achieved in two ways. The main bearings are ball and roller, and are situated on the same side of the crank throw such that they are immersed in a single oil bath in the sealed bearing housing. For the big and small end and piston rings, oil mist lubrication was used on a total loss basis. Two-stroke oil was supplied via a metal tube which protruded into the engine air inlet duct, the movement of air atomised the droplets of oil. A valve in the oil tube could be adjusted to regulate the supply, which was held in a burette so that its flow rate could be measured.

The ignition system was triggered by an optical sensor, and ignition timing could be easily varied. The spark plug washer was fitted with a thermocouple for temperature monitoring. The cylinder coolant circuit was arranged so that inlet and outlet temperatures could be monitored, and adjusted. Exhaust gas analysis equipment was available.

5 TEST PROCEDURE

The engine was tested at speeds of 1600, 1800, 2200, 2600 and 3000 rev/min, at both wide open throttle (WOT) and half open throttle (1/2 OT). At each speed and throttle setting, the air/fuel ratio was varied, and ignition timing adjusted to achieve maximum torque.

For each test point the following quantities were recorded:

engine speed (rev/min)
engine torque (Nm)
time for 1 cubic foot of gas to be used (sec)
manometer reading of pressure drop across
    orifice plate (mm H_2O).
ignition timing ($^\circ$BTDC)
spark plug temperature ($^\circ$C)
exhaust gas temperature ($^\circ$C)

The cylinder cooling water inlet temperature was maintained within the range 65-70 $^\circ$C. All the results were corrected to standard atmospheric conditions. For each test point, the power, thermal efficiency (both for net and gross calorific value), air/fuel ratio, scavenge ratio and brake mean effective pressure were calculated. The following results were generated graphically to show engine performance and tends:

(i)	BMEP	v Engine speed
(ii)	Power, torque	v Engine speed
(iii)	Delivery ratio	v Engine speed

(iv) Efficiency   v Engine speed
 (v) Peak exhaust temperature v Engine
     speed
(vi) Exhaust hydrocarbon concentration v
     Engine speed
(vii) Power, efficiency v Air/fuel ratio @
      1800 rev/min
viii) Power, efficiency v Air/fuel ratio @
      3000 rev/min

(delivery ratio is the mass of air delivered to the engine/displaced volume x ambient density)

For the sake of brevity, only the graphical results which are most significant are included in this paper. The above test procedure was repeated for a range of 5 inserts fitted to the gas inlet port, see Figure 1. This had the effect of changing gas port timing as follows:

Insert thickness (mm)	Gas port height (mm)	Gas port opening (^{o}ATDC)
0.0 (unmodified)	4.9	140.9
1.0	3.9	145.1
2.0	2.9	150.0
3.5	1.4	159.1
4.0	0.9	163.3
4.5	0.4	168.8

A full set of tests was carried out with each insert, and Figure 4 shows the maximum power and efficiency obtained for each insert. From these graphs it was concluded that the best gas port timing was 159.1 ^{o}ATDC (cf. with a gas port height of 1.4 mm). Figures 5 and 6 represent the engine perform-ance characteristics when the gas port was at its optimum timing.

6   SUMMARY OF TEST RESULTS

The maximum power obtained was 2.11 kW at 2,600 rev/min, and the maximum efficiency based on gross calorific value was 20.29% at 2,200 rev/min (22.47% on a net calorific value basis).

The effect of delaying gas port timing from the initial case where the gas port opened at the same time as the air transfer ports is that power does not change signifi-cantly, but efficiency increases by 15%.

7   METHANE METER

An infra-red gas analyser, which measures the percentage methane content of a sample of gas, was used to analyse gas samples taken from the crankcase and the exhaust pipe. Since natural gas is typically 92% methane, a figure for the quantity of fuel present in the sample can be determined.

The reading for the sample taken from the crankcase will determine the effectiveness of the stratification process. A value of zero would indicate perfect stratification, while a reading equivalent to the air/fuel ratio would mean that no stratification was occurring. The methane reading from the crankcase sample was negligible, and so it was concluded that perfect stratification of the air and fuel was indeed occurring.

Analysing the methane content of the exhaust gas will give an indication of the natural gas trapping efficiency of the cylinder scavenging process. The reading from the methane meter of the percentage of methane in the exhaust can be converted to the percentage loss of the fuel supplied to the engine through the exhaust port. This value, as well as being useful for comparing different porting arrangements, gives an indication of the potential for further efficiency gains if all the fuel charge can eventually be trapped.

Figure 7 shows the improvement in fuel trapping behaviour from the initial gas port timing, with no insert used, to the optimum gas port timing. The worst case is 45% fuel loss at 1600 rev/min, half-open throttle, with no insert in the gas port, and the best case is 18% fuel loss at 2,2000 rev/min, with the 3.5 mm insert in place.

If the 18% of fuel lost through the exhaust port could be trapped, it is estimated that the thermal efficiency would move to about 25% (gross CV basis) or 27.5% (net CV basis). This is an obvious area for continued development, and will be discussed later.

8   HOMOGENEOUS CHARGE

The engine could be operated, with appropriate modifications, in the conventional manner, where a homogeneous charge of air and gas is created in the crankcase. The engine was tested in this condition, so that the advantages of stratification (if any) could be quantified.

In summary, the results with homogeneous charge operation are not as good as the best stratified charge setting, but better than the worst. The highest gross efficiency obtained was 18.71% at 2,600 rev/min (cf. 20.29% at 2,200 rev/min for best stratified case, and 17.64% at 2,500 rev/min for worst stratified case i.e. no insert in the gas port).

9   FURTHER WORK

The next step in the development of this engine is intended to be the development of a fuel injection system. The gas is required to be introduced into the cylinder just before or just after exhaust port closure, in such a way that none can escape through the exhaust port.

The pressure in the cylinder at this juncture will be close to atmospheric, consequently the injector need only generate a pressure sufficient to move the gas charge into the cylinder. Problems arising from the time required for fuel evaporation, which complicate the design of gasoline injector systems, do not arise with a gaseous fuel. Bench tests are in progress on both a piston type and a diaphragm type injector, the next step is to incorporate one of them in the engine, and test it.

10   CONCLUDING REMARKS

A duty was specified for a natural gas fuelled heat engine drive for a heat pump. The engine specification is very similar to that for a cogeneration engine.

A crankcase compression two-stroke engine, with stratified charge, was designed, built, tested and refined.

The test thermal efficiency achieved was 20.29% (HHV basis), 22.47% (LHV basis).

At this efficiency the engine power output was 2.11 kW.

Further development is in progress designed to improve the thermal efficiency.

ACKNOWLEDGEMENTS

This project was funded by a British Gas/SERC co-funding grant, and British Gas have, in addition, been generous with equipment and advice.

Professor Haselden (University of Leeds) initiated the overall engine driven heat pump project, and his contribution is gratefully acknowledged.

Thanks are also due to the Department of Mechanical and Manufacturing Engineering, The Queen's University of Belfast, for the use of the excellent workshop and laboratory facilities.

REFERENCES

1. Kaizaki, M. 'Small gas engine heat pump systems'. 50th Autumn Meeting. The Institution of Gas Engineers. 1984.

2. Kuentscher, O.V. 'Mixture injection application for avoiding charge exchange losses in two-stroke cycle engines', Second Conference on the Small Combustion Engine, The Institution of Mechanical Engineers 1989.

3. Blair, G.P. et al 'A new piston design for a cross-scavenged two-stroke cycle engine with improved scavenging and combustion characteristics'. Transactions SAE 841096, 1984.

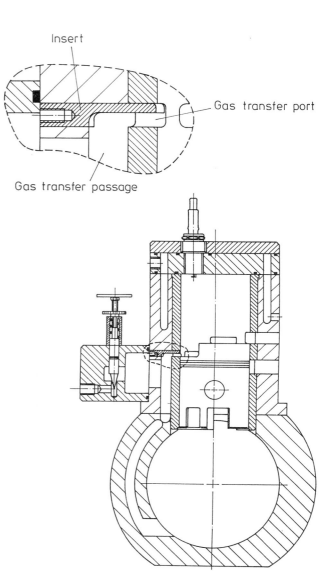

Fig 1   Location of timing insert

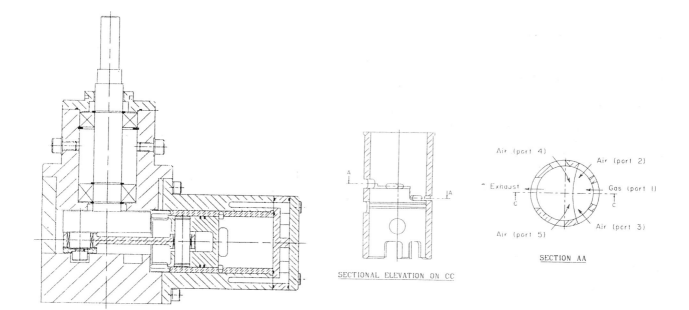

Fig 2   Port details

SECTIONAL ELEVATION ON CC

SECTION AA

Fig 3   Layout of main bearings

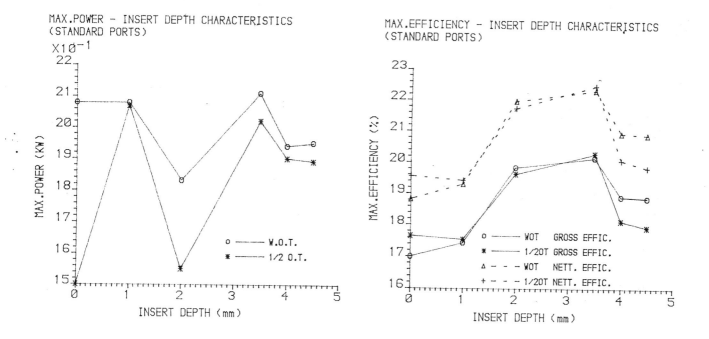

Fig 4   Maximum power and efficiency for each insert

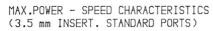

MAX.POWER – SPEED CHARACTERISTICS
(3.5 mm INSERT. STANDARD PORTS)

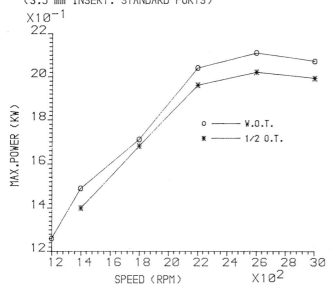

MAX.EFFICIENCY – SPEED CHARACTERISTICS
(3.5 mm INSERT. STANDARD PORTS)

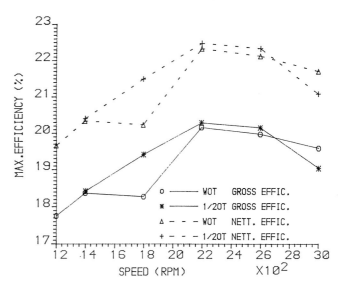

Fig 5   Test results

POWER – A/F RATIO CHARACTERISTICS @ 2200 RPM
(3.5 mm INSERT. STANDARD PORTS)

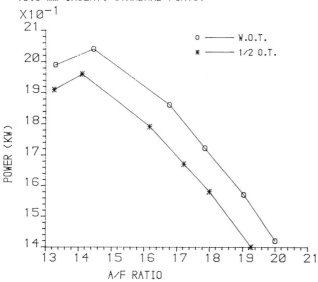

EFFICIENCY – A/F RATIO CHARACTERISTICS @ 2200 RPM
(3.5 mm INSERT. STANDARD PORTS)

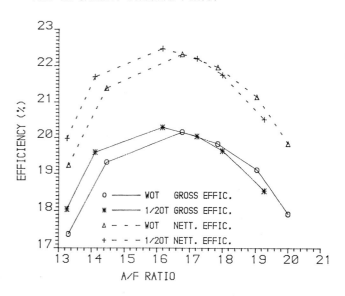

Fig 6   Test results

FUEL LOST AT MAX.EFFICIENCY – SPEED CHARACTERISTIC
(STANDARD PORTS. STRATIFIED CHARGE)

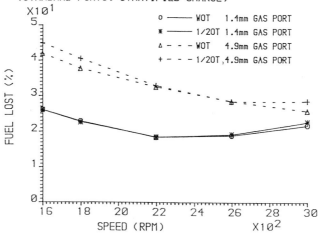

Fig 7   Fuel loss improvement

Oxygen control for gas engines with catalytic converters: a reliable low cost development

G R ROBERTS, BSc, MIMechE, MInstE, **R R THURSTON**, BSc, MInstP and **D J BROOMHALL**
British Gas plc, UK

SYNOPSIS Future emissions legislation may require the fitting of pollutant reduction equipment on industrial and commercial gas engine installations. This paper describes the development and successful testing of an exhaust oxygen control system on a 60 kW naturally aspirated gas engine in conjunction with a 3-way catalytic converter at the Midlands Research Station of British Gas. The system has been configured using commercially available control equipment and, in addition, a second version has been designed and tested for operation at high air/fuel ratios. These developments overcome the deficiencies of existing systems by providing accurate control of oxygen at low cost and with low maintenance requirements.

1 INTRODUCTION

The Midlands Research Station of British Gas has supported the potential application of natural gas fuelled engines to industrial and large commercial plant for a number of years. The work has been mainly aimed at ensuring that suitable engines are available and that the servicing requirements are minimal. Recently, concern over the environmental impact of fossil fuel combustion has added emphasis to the work aimed at quantifying gas engine emissions, and evaluating methods by which they might be minimised. In particular, the quantities of carbon dioxide, carbon monoxide, nitrogen oxides, sulphur dioxide and unburnt hydrocarbons are of interest.

The operation of small scale combined heat and power plant can make a worthwhile contribution to the reduction of carbon dioxide emissions by improving the overall efficiency of fossil fuel utilisation. In addition, operation on natural gas produces less carbon dioxide than any other hydrocarbon fuel and with no sulphur dioxide. However, careful control of engine combustion is necessary to prevent the formation of significant amounts of carbon monoxide or nitrogen oxides with both natural gas and other fuels.

This paper describes the design and testing of an exhaust emission control system, specifically for use on natural gas fuelled engines, which gives the twin benefits of low initial capital cost and reduced servicing requirements. The system was developed for operation at either stoichiometric or lean fuel/air ratios.

2 LEGISLATION

Legislation is already in force within several European countries to control the quantities of CO and NOx that can be emitted by engines, when used in automobiles and in industrial applications. Emissions reduction technology has been developed for vehicles and, separately, for large industrial plant. For technical and economic reasons, neither of these developments could be directly applied to small scale gas engine driven installations.

At the time of writing, there are no UK emissions standards with which installations should comply. However, it is expected that this situation will change with the introduction of statutory European Commission(EC) limits for gas, coal and oil fired plant. The proposed EC limits at present only apply to plant over 50 MW net thermal input, but limits for smaller installations will probably closely follow those imposed in West Germany and Holland, shown in Table 1.

Table 1 Emission limits for new plant - spark ignition engines (all sizes)

Component	Current	1992
a) Netherlands		
NOx (ppm at 3% O_2)	500	340
b) West Germany		
NOx (ppm at 5% O_2)	244	
CO (ppm at 5% O_2)	317	

3 EMISSIONS REDUCTION

There are two main methods by which NOx and CO engine emissions may be minimised. Firstly, operating a gas engine with high excess air will result in low NOx and CO levels, but with a rise in unburnt methane (Figure 1). The unburnt methane can be readily oxidised over a suitable catalyst. However, this so called lean burn technique results in reduced shaft power being produced, hence increasing the installed cost/kW of the installation. Power output may be increased back up to the original rating by the addition of a turbocharger to increase the charge density.

To take full advantage of the potential reduction in NOx levels using this technique, the excess air should be increased until it is very close to the maximum attainable before engine misfire occurs. The ratio of air being supplied to the engine compared with that required for stoichiometric combustion is defined as the lambda value. Control of the lambda value to within 0.05 of the setpoint would be acceptable. It should be noted that as weaker mixtures are used, the engine ignition timing will need to be advanced to maintain optimum combustion and power output. Advancing the ignition does not, however, maintain power output at the value obtained when running at stoichiometric ratio, but merely allows the optimum power at that air/fuel ratio to be attained.

Alternatively, the engine exhaust gases may be passed over a catalyst which simultaneously reduces the levels of CO, NOx and unburnt hydrocarbon. This is known as a three way catalyst and the two predominant reactions are:

$$2CO + O_2 \rightarrow 2CO_2$$
$$\text{and} \quad 2NO + 2CO \rightarrow N_2 + 2CO_2$$

Hence, The ratio of O_2 to CO in the exhaust gas is critical if NOx and CO are to be removed. If the ratio of O_2/CO is too high, there will be insufficient CO to reduce the NO and if the ratio of O_2/CO is too low, then significant concentrations of CO will pass through the catalyst. The air/fuel ratio must therefore be controlled over a very narrow range close to stoichiometric (1), as shown in Figure 2. The lambda value should be controlled to within 0.005 (0.5 percent at lambda of approximately 1.000) for correct catalyst operation.

As can be seen from the above, the requirements for the accuracy of control are stringent if satisfactory operation of the engine and reduction of emissions are to be ensured (2). The close control of air/fuel ratio must also be maintained between service intervals, without adjustments being necessary. This stability of control is vital for the small scale CHP units, where the cost of servicing must be kept to a minimum.

4 CONTROL SYSTEMS

Ideally the quantities of gas and air being supplied to an engine should be continuously monitored and the proportions adjusted to keep the ratio at the required level (stoichiometric, for operation with a three way catalyst). These proportions would vary as the ambient temperature, pressure and humidity varies from day to day. Unfortunately, low cost air and gas metering devices, with sufficient accuracy and reliability for industrial application, are not yet available, and so the direct measurement of air/fuel ratio is not easily carried out. However, for a given engine and ambient conditions, the combustion characteristics will remain fixed during extremely long periods of operation and the residual uncombusted oxygen level in the exhaust gases will be directly related to the air/fuel ratio of the mixture being supplied to the engine. The level of residual exhaust oxygen can therefore be used as the measured variable in a feedback control system.

The main advantage of using exhaust oxygen control is that changes in ambient temperature, pressure and humidity will be compensated for and the air/fuel ratio trimmed to the exact value on a continuous basis. Also, in situations where the calorific value of the fuel gas may vary, the required value of stoichiometric ratio will also vary and control of exhaust oxygen would automatically compensate for such changes, whereas open loop feedforward control of the air/fuel ratio would not.

4.1 Automotive control systems

Low cost air/fuel ratio control systems, based upon the measurement of residual exhaust oxygen, have been used for many years for automotive applications, where three way catalysts have been employed. Zirconia oxygen sensors have been developed (3) for the severe operating environment found on vehicles. The reliability and repeatability of these sensors have been improved by the addition of an integral heater element to maintain the cell temperature above its operating minimum (4). The temperature of the cell is, however, not controlled to any particular value and is dependent upon a number of factors (exhaust gas temperature, vehicle battery voltage, position in exhaust system).

The automotive control systems (1) are operated on a limit cycle by varying the induction mixture strength from rich to lean over a period of about one second and observing the change in zirconia cell output. Figure 3 indicates that the cell output varies dramatically as the air/fuel ratio is varied either side of stoichiometric. Operation of gasoline engines with three way catalysts requires a stoichiometric ratio of unity and such an oscillating control system produces a varying air/fuel ratio with an average of unity. These systems also oscillate at a rate determined by the dynamics of the fuel system, induction manifold, engine and exhaust system, which for a vehicle, tend to be fast acting.

Fuelling on natural gas requires a slightly rich setting for correct catalyst operation (2). A limit cycle would thus result in severe rich fuelling excursions at one extreme of the zirconia cell characteristic. This, coupled with the slower speed of response of industrial gas fuelled engines means that a modified version of an automotive controller would not be suitable as it would result in a very slow limit cycle with substantial periods outside the required catalyst operating window.

4.2 Industrial control systems

Exhaust oxygen control systems, such as those developed by Ruhrgas (5) and Waukesha, amongst others, have been developed for the larger sizes of industrial gas fuelled engines. To use one of these systems on a small scale gas engine would increase the capital cost significantly and also lead to an increase in the maintenance requirements. The high cost is partly caused by the need to apply temperature correction to the zirconia cell output voltage in order to measure the oxygen concentration accurately. In addition, the reliability of the zirconia cells have been variable, between 500 and 2500 hours of operation being reported as the life expectancy. Cell failure due to ageing (4) is believed to be caused by high exhaust temperatures and thermal shock due to variations in temperature during engine load changes.

5 DEVELOPMENTS AT BRITISH GAS

Over the last few years, developments of furnace monitoring equipment at British Gas have used the heated zirconia cell manufactured by Bosch (4) in an oxygen analyser (6). This system uses temperature control of the cell heater and a digital technique for linearising the millivolt output of the cell so that the final voltage is directly proportional to oxygen concentration. Reliability has been excellent, with continuous operation in excess of 2 years being reported without failure or cell replacement.

The engine control system described within this paper is based upon the application of the furnace oxygen monitor, together with a two term, proportional and integral (PI), process controller and an electrically adjustable gas supply pressure regulator, as shown in Figure 4. The work so far undertaken at the Midlands Research Station has been on engines fitted with the 'IMPCO' gas carburettor and supplied with natural gas at a pressure of 20 mbar. The standard gas carburettor is retained to carry out the primary gas/air ratio control and mixing, with the carburettor gas inlet pressure being varied, as required by the oxygen signal, in order to trim the ratio to that required.

5.1 Lean - burn control system

This system uses a standard oxygen monitor as shown in Figure 4, where the monitor produces a 0 - 2.1 volt output signal representing an oxygen level of 0 - 21%. This output voltage is passed to the PI controller, where it is compared with the voltage setpoint and a dc milliamp current produced to open the pressure regulator as necessary. The required voltage setpoint is determined from the characteristic in Figure 5 for any particular excess air level.

5.2 Stoichiometric control system

Figure 6 shows the configuration of the control system for operation at close to the stoichiometric ratio, where the zirconia cell produces a much higher millivolt level which is applied directly to the PI controller, without the necessity of signal amplification or linearisation. In this case, the oxygen monitor unit is used solely to provide the cell heater current/temperature control circuit. The required voltage setpoint for the PI controller can be determined from the zirconia cell characteristic shown in Figure 3.

6 CONTROL SYSTEM TEST WORK

Testing has been carried out on three different engines during the development of the control systems. The engines were a Waukesha VRG220 4 cylinder, 3.6 litre, a Ford SI4, 4 cylinder, 4.12 litre and a Ford SI6, 6 cylinder, 6.18 litre.

6.1 Emissions reduction results using the lean burn system

The lean burn control system was tested on a Waukesha VRG220 gas engine. These tests were carried out partly to determine the emissions reduction which could be obtained from lean fuelling and partly to prove the control stability of the system. An oxidation catalyst was not fitted to the exhaust.

Table 2 shows the exhaust emission levels of CO, CH_4, and NOx, with the oxygen control system adjusted to give excess air levels close to the engine misfire limit. These results clearly demonstrate the potential reduction in NOx emissions when using high excess air levels. However, further work will be required in order to extend engine misfire limits and examine the performance of oxidation catalysts to reduce the level of methane emitted.

Table 2 Exhaust emissions from Waukesha VRG220 (ignition timing optimised)

Air/Fuel ratio	NOx vpm	CO %v/v	CH_4 %v/v
1.045	>4000	0.040	0.085
1.430	235	0.050	0.165

6.2 Emissions reduction results using the stoichiometric system

The stoichiometric system was tested on a Ford SI6 gas engine, with the exhaust system fitted with three 3-way catalyst monoliths in series, these being supplied by Johnson Matthey, type P12CEB/50/5.1. This series arrangement was used to determine how much catalyst surface

Table 3 Analytical techniques and equipment

Component	Technique	Make and Range of Equipment
NOx	Chemiluminescence	Analysis Automation Model 443 0 - 10, 25, 100, 250, 1000, 2500 and 10 000 vpm.
CO	Non-dispersive infra-red spectroscopy	Analytical Development Co., Ltd. RF Series 0 - 0.1, 1 and 5% v/v.
O_2	Paramagnetism	Servomex Model 570A 0 - 100 % v/v.
CH_4	Non-dispersive infra-red spectroscopy	Analytical Development Co., Ltd. RF Series 0 - 1, 5 and 10 % v/v.
CO_2	Non-dispersive infra-red spectroscopy	Analytical Development Co., Ltd. RF Series 0 - 1, 10 and 15 % v/v.
H_2	Gas Chromatography	Philips Scientific PU4500 fitted with thermal conductivity detector. 0 - 100 % v/v, depending on calibration gas.
N_2O	Fourier-transform infra-red spectroscopy	Nicolet SBX20 fitted with Spectra-Tech 10m. path length gas cell. Linearity established for 0 - 25 vpm.
NH_3	Ion selective electrode	Russell 9512 10^{-4} to 10^{-2} molar.

Table 4 a) Exhaust component concentrations at 19 hours (stoichiometric control at setpoint of 780mv)

Component Concentration	Before Catalyst Monolith	After 1st Catalyst Monolith	After 2nd Catalyst Monolith	After 3rd Catalyst
NOx (vpm)	2600	120	<10	<10
CO (% v/v)	1.15	0.03	0	0
CH_4 (% v/v)	0.18	0.10	0.09	0.08
O_2 (% v/v)	0.8	0	0	0
CO_2 (% v/v)	10.8	12.4	12.4	12.4

b) Exhaust component concentrations at 4884 hours (stoichiometric control at setpoint of 780mv)

Component Concentration	Before Catalyst	After 1st Catalyst Monolith	After 2nd Catalyst Monolith	After 3rd Catalyst Monolith
NOx (vpm)	2690	75	12	9
CO (% v/v)	1.25	0.012	0	0
CH_4 (% v/v)	0.11	0.05	0.04	0.04
O_2 (% v/v)	1.1	0	0	0
CO_2 (% v/v)	10.7	12.2	12.1	12.1

area would be required for this particular engine, with emissions measurements being made before and after each monolith. In addition, the long term durability of the catalysts and the stability of the control system were studied, with the engine operating at 1500 r/min, wide open throttle. At the time of writing, the engine has run continuously with the minimum of interruptions for almost 5000 hours.

Comprehensive gas analysis equipment was used, the details of all analytical techniques and equipment being given in Table 3.

The concentrations of NOx, CO, CH_4, CO_2 and O_2 before and after each catalyst monolith are listed in Table 4. These results indicate that NOx and CO levels have been reduced to well below the current European limits after the second catalyst monolith and that the third unit is not necessary for the volume of exhaust gas flow from this particular engine. Although not subjected to any emissions limits at the present time, the values of unburnt methane before and after the catalysts have shown that a 50 percent reduction takes place.

Throughout the long term test duration, the zirconia cell control setpoint has not been adjusted from the original value of 780 millivolts, indicating the minimal servicing requirements of the control system.

Previous workers (2) in the field of emission control have indicated that the three way catalyst may produce ammonia, which would obviously be environmentally unsatisfactory. In order to determine whether either ammonia or nitrous oxide were produced in the exhaust, comprehensive gas analyses were carried out. The results of these tests are shown in Table 5 and indicate only extremely low levels are present.

6.4 Control system response tests

In order that the speed of response of the oxygen control systems could be examined, some basic test work was carried out with a Ford SI4 engine. This work was done with a redesigned control system which could be readily reconfigured for either stoichiometric or lean burn operation. In order to cause the maximum change in gas flow during these tests, the speed and load were simultaneously altered from minimum (idle) to maximum (run) and back again. In addition, for the lean burn option, the tests were repeated at two values of control setpoint.

The results of these tests are shown in Figures 7, 8, and 9, the settling times for the stoichiometric and lean burn systems were 30 and 25 seconds respectively. It is worthwhile pointing out that these tests were more severe than would be experienced in practice, where the engine speed would generally be kept constant and the load varied.

7 FUTURE WORK

It is planned to determine the long term reliability of the lean burn option and investigate ways of reducing the levels of unburnt methane in the exhaust gas when using the stoichiometric system. It is also intended to examine the potential for further reductions in system cost by integration of the control function and the zirconia cell heater function into one electronic unit.

Although the system has been tested with small scale engines, it should operate equally well with larger power plant. The identification of gas flow control valves suitable for higher gas flow rates will be necessary.

Table 5 Exhaust component concentrations measured at 4864 hours

Sampling Point	NOx vpm	Component concentration (dry basis except for NH3)						
		CO % v/v	O_2 % v/v	CH_4 % v/v	CO_2 % v/v	H_2 % v/v	N_2O vpm	NH_3 vpm
Before catalyst	2600	1.23	1.1	0.11	11.2	0.76	<5	<5
After 1st catalyst monolith	55	0.012	<0.1	0.06	12.7	0*	<5	<5

* H_2 detection limit 0.05% v/v

6.3 Accuracy of control

Both the lean burn and the stoichiometric control systems were initially tested on a Waukesha VRG220 gas engine in order to determine the accuracy to which a particular setpoint could be controlled. It was found that the lean burn system could control the oxygen level to within 0.2 per cent of the required value, this being equivalent to a variation of lambda ratio of 0.02 at a setpoint of 1.5. The stoichiometric control could maintain the lambda ratio to within 0.004 at a nominal setpoint of 1.000.

8 CONCLUSIONS

An exhaust oxygen control system has been designed for use on small scale gas engine installations, where the available gas control valves limit the application to engines producing a maximum of about 300 kW shaft power. The system can be readily re-configured for either stoichiometric or lean burn applications. The stability and response testing indicated that satisfactory operation would be obtained for the essentially constant engine speeds and loads encountered during industrial and commercial duty. Long term durability testing has proven that emissions can be consistently reduced to below current European limits with excellent reliability, no adjustments being necessary over almost 5000 hours of operation.

9 ACKNOWLEDGEMENT

The authors would like to thank British Gas plc for permission to publish this paper.

10 REFERENCES

(1) Bosch Technical Instruction. Emission control for spark ignition engines.

(2) KLIMSTRA, J. Catalytic converters for natural gas fuelled engines - a measurement and control problem. SAE paper 872165, November 1987.

(3) Bosch Technical Instruction. Exhaust gas lambda sensor.

(4) WIEDENMANN, H.M., RAFF, L., NOACK, R. Heated zirconia oxygen sensor for stoichiometric and lean air-fuel ratios. SAE paper 840141, February 1984.

(5) KORSMEIER, W., WOLF, D. Homix system. Gas mixing unit for gas engines. BWK 1989 No. 6/7.

(6) British Gas, Midlands Research Station, Relay magazine. Oxygen monitor now available, pp 11, September 1987.

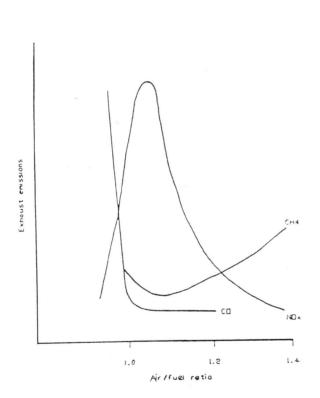

Fig 1   Gas engine exhaust emissions

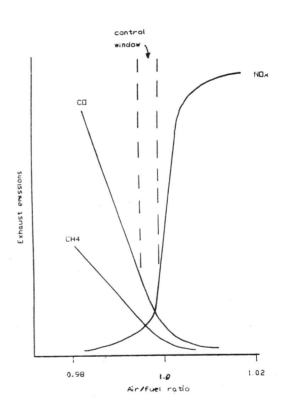

Fig 2   Emissions after three way catalyst

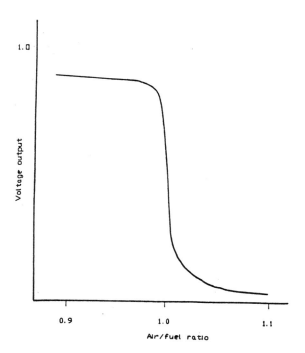

Fig 3   Zirconia cell characteristic

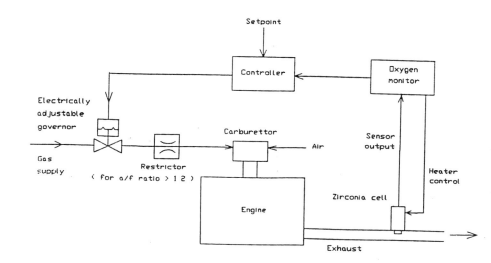

Fig 4   Lean burn control scheme

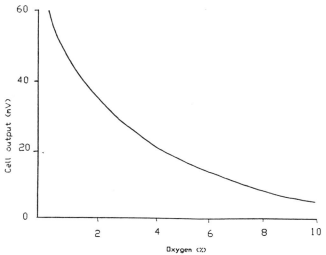

Fig 5   Zirconia cell characteristic for operation at high
excess air levels

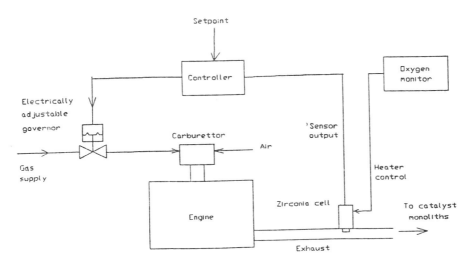

Fig 6   Stoichiometric control scheme

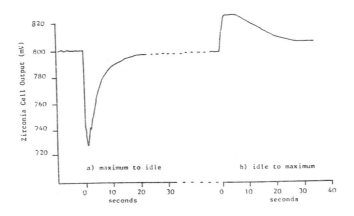

Fig 7   System response (stoichiometric)

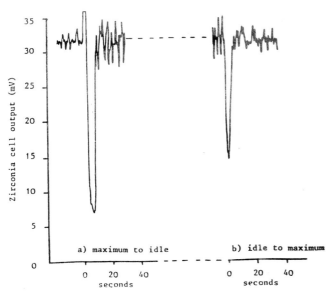

Fig 8   System response (lean, 2% oxygen)

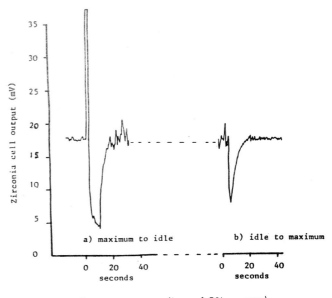

Fig 9   System response (lean, 4.3% oxygen)

Recent developments in gas/air mixers and in microprocessor air/fuel ratio control systems

P TIEDEMA and **L WOLTERS**
Deltec Fuel Systems BV, The Netherlands

SYNOPSIS In this paper, recent developments in relation to the DELTEC gas carburettors and air/fuel ratio control systems are described. The most important features of a gas/air mixer are as follows: Air/fuel ratio; mixture homogeneity; pressure drop, and reliability. Attention is given to the gas/air mixer-pressure regulator system. For turbocharged gas engines mixture boosting is compared to mixing downstream of the turbocharger. Microprocessor air/fuel ratio control systems for stoichiometric and for lean burn gas engines are of essential value in meeting and maintaining stringent maximum exhaust emission levels. Attention is given to their potential in controlling air/fuel ratio under varying input conditions. Operational experience is reported.

1 INTRODUCTION

Gas engines are increasingly being used for co-generation applications. In countries where natural gas pipe lines form a dense grid, such as in The Netherlands and West-Germany, the application of combined heat and power generation using natural gas fuelled engines is increasing.
In The Netherlands by the end of 1989 the total number of gas engine driven co-generation plants was approx. 2500 with a total installed power of 300 MW. For the next years an increase is expected with 100 MW per annum.

In view of the increasing demands based on energy and environmental considerations, the technical requirements on gas engines are becoming more and more severe. This affects areas such as preparation and metering of the gas/air mixture, design of the intake system and combustion chamber, the ignition system and the use of engine monitoring.

Deltec Fuel Systems developes and manufactures fuel systems for various applications with emphasis on heavy duty gas engines. This paper describes some of the DELTEC products, in particular the gas carburettors and the fuel control systems. Performance in view of energy efficiency and achievable exhaust emission levels are discussed.

2 GAS CARBURETTORS

The preparation of the correct gas/air mixture is of vital importance for optimum combustion, good engine performance and low exhaust emissions. A well designed gas carburettor provides the possibility to meet strin-

gent requirements in this respect. Instead of using a carburettor, gas can also be supplied to the engine by means of an injection system. However, this requires higher pressure than is usually available and, moreover, mixture homogeneity will not be as good (1). For stationary dual fuel engines gas injection is normally applied (2,3).

2.1 Operating principle, air/fuel ratio

Deltec has developed a range of gas carburettors for power outputs in the range 30-2000 kW. These carburettors are based on the venturi principle, and they have been designed to meet the following requirements:
* constant air/fuel ratio at varying mixture flow
* high homogeneity of the gas/air mixture
* low pressure drop over the venturi
* fast and accurate response to changes in engine speed and load
* easy installation and adjustment
* maintenance free operation

The operating principle of the DELTEC carburettor is described in the following section. Fig 1 illustrates the basic lay-out.
The air needed for combustion passes the air filter and enters the venturi. The smaller diameter of the venturi throat results in an increase in air velocity and a drop in pressure. If incompressibility of the flow is assumed the relationship between flow and pres-

sure drop can be described with Bernouilli's law as follows:

$$\Delta p_a = 1/2 \, \rho_a \cdot v_a^2$$

(a = air)

The pressure drop over the restriction in the gas supply line is created by the combination of the main adjustment screw and the gas metering orifices in the venturi throat.

$$\Delta p_g = 1/2 \, \rho_g \cdot v_g^2$$

(g=gas)

When Δp_a equals Δp_g which is effected by means of the zero-pressure regulator, then a linear relationship exists between v_a and v_g.

The mass flow of air through the venturi is:

$$\dot{m}_a = \rho_a \cdot v_a \cdot A_a$$

$$\Delta p_a = 1/2 \, \rho_a \cdot v_a^2$$

$$\dot{m}_a = \rho_a \cdot \sqrt{\frac{2\Delta p_a}{\rho_a}} \cdot A_a$$

where A_a = area of venturi throat.

Similarly the mass flow of gas through the venturi throat orifices is:

$$\dot{m}_g = \rho_g \cdot \sqrt{\frac{2\Delta p_g}{\rho_g}} \cdot A_g$$

where A_g = area of venturi orifices.

In the venturi throat where air and gas are mixed the pressure drop of air flow and of gas flow are equal:

$$\Delta p_a = \Delta p_g$$

The air/fuel ratio is obtained by the quotient of the mass flows of air and gas:

$$\frac{\dot{m}_a}{\dot{m}_g} = \frac{A_a}{A_g} \cdot \sqrt{\frac{\rho_a}{\rho_g}}$$

Using the assumption that the density of air and of gas are constant, the air/fuel ratio is constant. In reality this is not the case, since density varies with humidity, temperature and pressure. If however the velocity pressure is small in relation with total pressure the assumption of constant density is reasonable. Furthermore, in many cases gas composition fluctuates. It depends on the emission and/or detonation requirements if additional air/fuel ratio control is necessary to eliminate the forementioned effects. As the air filter becomes dirty during operation, the increase in air flow restriction will result in a larger pressure drop and consequently in enrichment of the gas/air mixture. Therefore the use of a balance line between air filter and zero pressure-regulator is required to compensate for this effect.

The air/fuel ratio required for proper combustion depends on the engine and its mode of operation. For stoichiometric running engines the air/fuel ratio should be constant over the complete speed and load range. Lean burn engines require an air/fuel ratio which is high at full load to ensure that NOx emission is kept as low as possible. At partial load the gas/air mixture has to be somewhat richer to avoid misfiring. At starting the mixture should be close to stoichiometric. The carburettor has to meet these basic requirements as best as possible. This is also important in the case of engines which use separate air/fuel ratio control. Such systems often operate with feed back signals from engine exhaust sensors, but this should be for fine tuning rather than the basic air/fuel ratio control.

The correct combination of carburettor and zero pressure-regulator is essential in maintaining and controlling the correct air/fuel ratio over the whole engine speed and load range. In practice there are limitations in accuracy, mainly because of hysteresis in the pressure regulator. With proper dimensioning of the gas metering area the DELTEC carburettor provides sufficient accuracy. In that respect it is also important that the restriction created by main adjustment screw and metering orifices is such that approximately 80% of the total restriction is over the orifices and only 20% over the main adjustment screw. Fig 2 illustrates the effect of different settings.

The required basic air/fuel ratio can be adjusted by means of the main adjustment screw between the zero-pressure regulator and the carburettor. The DELTEC carburettor has a modular design, which includes the main adjustment screw. This component can very easily be exchanged for a pneumatic or electronic air/fuel ratio control in case separate lambda control is required.

2.2 Homogeneity

For stable running of the engine and low exhaust emissions homogeneity of the gas/air mixture is essential. The DELTEC carburettor can achieve very high homogeneity with local lambda variations within a 1% tolerance. This is achieved by mixing gas and air not only at the circumference of the venturi throat but also over the cross section of the venturi throat through a spike or cross. It is very important to avoid moving parts in the gas/air mixing area. Because of maladjustment or wear of such parts homogeneity can deteriorate at a dramatic rate.

Tests (4) indicate that homogeneity may be improved if a certain additional mixing length between the venturi and the throttle valve is used. The length of this should be in the order of magnitude of 5 D, where D is the diameter of the venturi throat. Although the effect seems obvious in a positive sense, field experience does not always support the need for this. Moreover space for such mixing tubes is not always available and furthermore there has to be a compromise between higher homogeneity and increased flow resistance. For turbocharged engines with the carburettor upstream of the turbocharger such additional mixing length is certainly not necessary.

2.3 Pressure drop

The pressure p_g in the venturi throat is the controlling factor for the amount of gas flowing into the venturi, and is thus called the gas signal pressure. For starting the engine this signal needs to be above a minimum value and therefore a certain minimum air velocity is necessary. On the other hand the lower the air velocity the lower the resistance over the carburettor, which is important for power output, fuel consumption and NOx.

The low resistance of the DELTEC carburettor has been achieved by realizing an almost ideal droplet shape of venturi and spike or cross. Since the dimensions can be adapted to a specific engine, the optimum lay out for each engine type and power output is possible. The peak mixture flow is usually established at 80m/sec for stationary engines, which results in a pressure drop of approx. 3,5 kPa. This should however still give sufficient signal for starting conditions.

The ever increasing difference in mixture flow at idle speed and at full load / high speed, being caused by reducing idle speeds and increasing boost pressures, may introduce the necessity for an optimum setting at full load and the use of a separate system for good starting performance.

2.4 Pressure amplifier control system

For engines where ultra low resistance is required, a so-called pressure amplifier control system has been developed. This system makes use of the available inlet pressure. Therefore the suction in the venturi throat to create suffi-

cient gas signal pressure drop is lower than without this system. Consequently the resistance can be lower.

Fig 3 shows the basic lay-out. The working principle is as follows. The pressure regulator is modified so that instead of an internal pressure feed back line an external pressure feed back is available. This external pressure feed back line is connected to a pneumatic potentiometer. This potentiometer receives two signals, one from a pressure sensing tube in the venturi throat, the other from the gas inlet chamber at the circumference of the gasmetering orifices in the venturi throat. As soon as there is an air flow through the venturi it results in a negative pressure in the pressure sensing tube in the venturi throat. The location of this tube in the middle of the air flow gives a more sensitive signal than that from the metering orifices. The negative pressure will open the pressure regulator. Subsequently the outlet pressure of the regulator (positive pressure) will increase until it equals the negative pressure. When the resistors in the pneumatic potentiometer have the same resistance the signal in the pressure feed back line then becomes zero and the system is in balance. In case a leaner gas/air mixture is desired the resistance in the positive pressure line must be decreased with at the same time that in the negative pressure line is increased. For obtaining a richer mixture, the resistances change the other way round.

When an air/fuel ratio control system is used or when a varying gas quality requires a variable resistance in the main gas flow this system is of great use. As it controls the main gas flow by means of the pressure feed back line on the regulator, there is no need for powerful accurate expensive high capacity actuactors in the main gas line. This is especially important for larger installations. Deltec applies this system standard on its lean burn engine air/fuel ratio control system.

2.5 Turbocharged engines

Turbocharging is increasingly being applied to gas engines, certainly in the higher power range. DELTEC gas/air mixers can be installed upstream of the turbocharger. There is no problem with surge of the compressor because of the low resistance of the carburettor. This way mixture homogeneity is enhanced by the turbocharger.

For large turbocharged engines with air-to-air intercooling installation of the mixing unit down-stream the turbocharger is preferred, to avoid a large amount of ignitable air/gas mixture. This way of gas supply requires balancing of the output pressure of the pressure regulator with the boost pressure of the turbocharger. Fig 4 shows the lay-out of such installation. It should be noted that it is important to have available an inlet gas pressure that is at least equal to the boost pressure. To compress gas from approx. 30 mbar to 2000 mbar will cost approx. 5% of the output power of the engine.

3. AIR/FUEL RATIO CONTROL SYSTEMS

In several countries the maximum allowed exhaust gas emission levels will be further reduced. Moreover, in some cases the user receives financial incentives if lower emission levels than maximum allowed can be demonstrated and maintained during service life of the installation. In West-Germany several users already require emission levels not surpassing 50% of the regulated maximum exhaust emission levels.

It is not possible to meet such emission requirements without the use of advanced electronic control systems. The basic setting of the standard carburettor-pressure regulator may be capable to meet the specifications initially,but fluctuations in gas quality, variation in ambient conditions and wear in system components will lead to deterioration such that at in time emission levels cannot be maintained anymore. Modern microprocessor air/fuel ratio control systems enable the user to meet and to maintain maximum permitted emission levels.

Deltec Fuel Systems has developed such systems for stoichiometric operated engines with 3-way catalyst as well as for lean burn engines with oxydation catalyst. Both systems make use of standard basic components to a large extent.

3.1 Microprocessor air/fuel ratio control system for lean burn engines

Lean burn gas engines are used when high mechanical efficiency is required in combination with low exhaust emissions. The relatively high amount of excess air results in much lower NOx formation during combustion than with stoichoimetric operated engines. The air-excess ratio is limited by the misfiring borderline. This results in a very narrow operation range for

the lean burn engine, as shown in Fig 5. The main advantage of lean burn operation is the higher efficiency, emissions generally being somewhat higher than those of stoichiometric running engines with 3-way catalyst, especially HC-emissions. To minimise these, an oxydation catalyst is used. Fig 6 shows the basic closed loop control system which consists of:

* electronic control module (ECM)
* sensors for inlet manifold pressure and temperature, generator power, engine speed irregularity
* lean exhaust oxygen sensor
* stepper motor controlled gas valve
* power supply

This assembly is completed with a pressure regulator and a gas carburettor.

The control system can simply be monitored with a personal computer, which shows:

* measured air/fuel ratio
* desired air/fuel ratio
* calculated air/fuel ratio
* inlet manifold absolute temperature
* inlet manifold absolute pressure
* gas valve position
* generator power
* malfunctions

Remote monitoring by means of a modem is also possible.

The electronic control module is a multi-function engine management system computer with self diagnosis which monitors the sensors, connections, internal electronics and operating conditions to assure correct operation. Predefined in the internal memory of this ECM is a table of optimal air to fuel ratios as a function of generated power. From known engine characteristics and the sensor signals the mixture flow and the actual air to fuel ratio are calculated. The ECM will always activate the gas metering valve in such a way that the desired air to fuel ratio is maintained. The generator power output, which is used as feedback signal, enables the system to cope with varying gas qualities and changes or malfunctions in the gas supply system.

However, generator output power alone would be not sufficient as a feed back signal. This is because all causes of power loss, such as misfiring, loss of engine efficiency, mechanical defects etc. will be corrected by enrichening the air/fuel mixture. This enrichment will be kept within safe limits by the ECM, but the addition of other sensors, primarily the lean oxygen sensor and the PASTOR enables operation much closer to the safe operation limits of the engine, thus widening the control range of the system.

With this complete control system, sensor calibration changes can be detected as well, since more sensors provide information from which engine related as well as gas system and sensor related problems can be discerned. The system then can distinguish power changes due to gas supply and those due to other causes and it will only correct for changes in gas supply and gas quality. The other causes, which will be engine or generator related, will result in a malfunction code followed by an alarm and ultimately engine shut down.

The lean oxygen sensor is essential for direct feed-back to the ECM of the air/fuel ratio. It has a linear and sufficiently steep characteristic from lambda = 1.2 up to lambda = 1.7 as shown in Fig 7.

The control system is standard supplied with the PASTOR (Prime mover Angular speed Stability moniTOR). It protects the engine from damages caused by improper functioning of cylinders, e.g. when misfiring occurs. The unit measures at a high accuracy level the variation in angular speed during each crank shaft revolution. When the measured non-uniformity level exceeds a certain set value an alarm is activated. At further increased or at prolonged non-uniformity level a second adjustable level detector automatically commands engine shut-down. The PASTOR is so sensitive that also in case the generator is coupled to the electricity grid it detects speed irregularity. The PASTOR output signal is an input signal for the ECM.

Corrections to the air/fuel ratio are made through the action of a stepper motor control valve. This valve controls a pressure divider, which receives two pressure signals, one upstream the main adjustment screw, the other coming from the venturi throat. The differential pressure which is continuously adjusted by the stepper motor valve is connected to the zero pressure regulator. This system enables very accurate metering of gas in feed back control.

3.2 Microprocessor air/fuel ratio control system for stoichiometric engines

Gas engines equiped with 3-way catalyst are used in case the lowest possible emissions are required. The 3-way catalyst reduces and oxidizes in only a very narrow window around the stoichiometric air/fuel ratio. A slightly rich mixture will prevent the oxidation of

CO and HC while a slightly lean mixture will do the same regarding the reduction of NOx, as shown in Fig 8.

A venturi type gas/air mixing unit will basically provide a constant air/fuel ratio at all engine speeds and loads. However, for a 3-way catalyst operating at its highest conversion rate an uncontrolled mixing unit will not maintain the air/fuel ratio sufficiently accurate. This is caused by imperfect operation of pressure regulators, varying air and gas temperature etc. A sophisticated control system is therefore necessary.

The control system can be simpler than that for lean burn engines. It consists of the ECM, an exhaust oxygen sensor and an air/fuel ratio control valve. This assembly is combined with a gas carburettor and a zero-pressure regulator. The system enables open loop operation for, for example, when the engine is cold, whilst providing closed loop operation as soon as normal operating conditions exist.

4 FIELD EXPERIENCE

DELTEC carburettors are in use on many different engine types and makes. Table 1 shows emission results obtained in lean burn operation. The first engine is equiped with a lean burn air/fuel ratio control system and oxydation catalyst. In brackets the emissions values after the catalyst have been given.

Table 1 Emission results with lean burn engines

Engine/speed/power rpm (kWe)	Lambda	NOx mg/Nm^3	CO mg/Nm^3
CAT 3412 TA/1500/360	1,53	333 (310)	443 (0)
CAT 3516 TA/1500/1050	1,58	297	630
MAN 2842 NA/1500/165	1,58	284	581
Emission limits TA-Luft:		500	650

With stoichiometric operation very low emisions can be obtained. Table 2 shows emissions after the 3-way catalyst.

Table 2 Emission results with stoichiometric engines with 3-way catalyst

Engine/speed/power (rpm) (kWe)	NOx mg/Nm^3	CO mg/Nm^3
MAN 2842 NA/1500/220	40	217
MAN 2866 NA/1500/110	40	217
CAT 3408 TA/1500/360	68	232
Emission limits TA Luft:	500	650

The DELTEC lean burn air/fuel ratio control system has been installed on engines which run on landfill gas with large variations in CH_4 content. The system automatically corrects for fluctuations between 28-50% CH_4. Table 3 shows some results.

Table 3 Emission results with DELTEC lean burn control system on engine running on landfill gas

Caterpillar 3516/1500 rpm/500 kWe

	lambda	NOx mg/Nm³	CO mg/Nm³
28% CH_4	1,50	315	487
45% CH_4	1,52	470	551

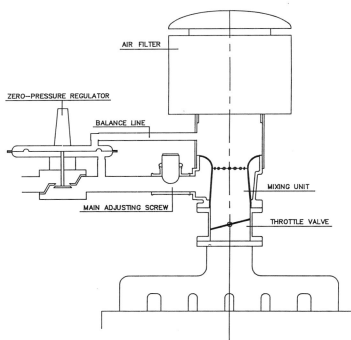

Fig 1   Concept of Deltec venturi carburettor with zero pressure regulator

REFERENCES

(1) HUNDLEBY,G. Low emissions approaches for heavy duty gas-powered urban vehciles. SAE paper 892134, International Fuels and Lubricants Meeting and Exposition, Baltimore, 1989.

(2) GRONE,O. and PEDERSEN,P.S. Large diesel engines using high pressure gas injection technology. Gasfired co-generation power plants, Aarhus, 1989.

(3) VESTERGREN,R. High pressure gas engines. Gasfired co-generation power plants, Aarhus, 1989.

(4) KLIMSTRA,J.Carburettors for gaseous fuels - on air-to-fuel ratio, homogeneity and flow restriction. SAE-paper 892141, International Fuels and Lubricants Meeting and Exposition, Baltimore, 1989.

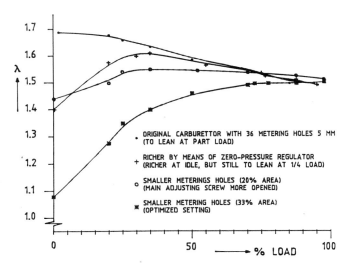

OPTIMALISATION OF A CATERPILLAR 3516 GASENGINE EQUIPPED WITH TWO DELTEC 200-II CARBURETTORS

Fig 2   Effect of variation in restriction over gas metering orifices on air/fuel ratio curve

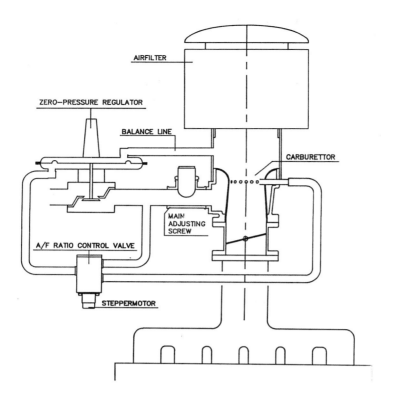

Fig 3   Deltec pressure amplifier control system

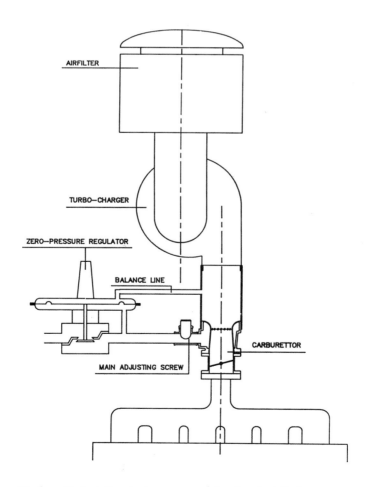

Fig 4   Carburation system was gas/air mixer installed
downstream the turbo charger

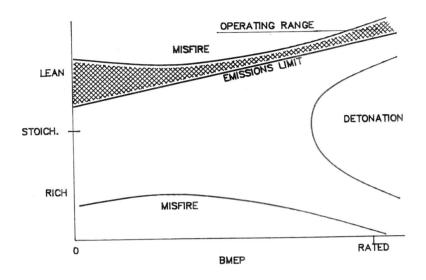

Fig 5   Operating range of lean burn gas engines

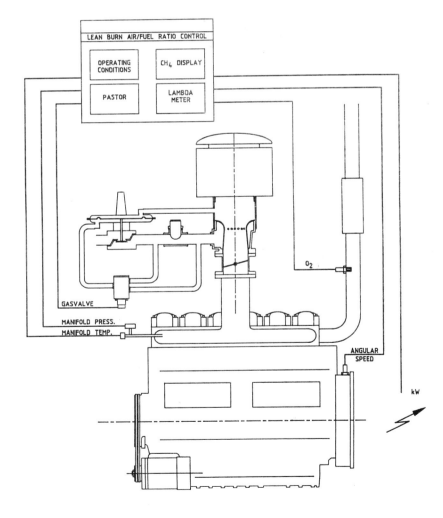

Fig 6   Deltec microprocessor air/fuel ratio control system
for lean burn gas engines

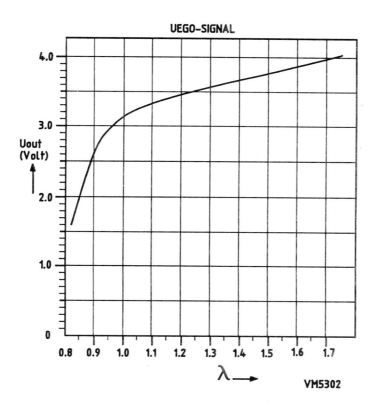

Fig 7   Characteristics of UEGO lean oxygen sensor

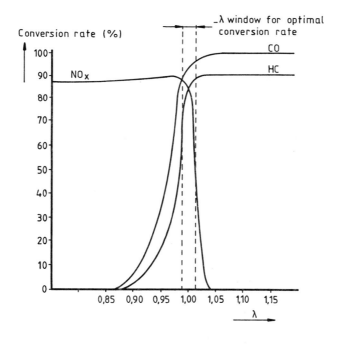

Fig 8   Lambda window for optimal catalyst conversion
          rate

The lubrication of gas engines

R J GILBERT, CEng, FIMarE and **D B WOOTTON**, AMIRTE, FIDiagE
Mobil Oil Co Limited, UK

The advent of the spark ignited gas engine in preference to its older sister, the duel fuel engine has brought about the need for improved understanding of the very specific role modern gas engine lubricants have to play. This paper discusses the lubrication of gas engines and the associated problems arising from the use of differing gases as fuels. It looks at the use of biogases arising from landfill sites or sewage digester plants and examines some of the mechanisms of both corrosion and deposition which may have an adverse effect on both the life of the engine and its lubricant charge.

We will discuss how the combination of used lubricant analysis and the borescopic examination of the combustion spaces can lead to a better understanding of the corrosion/wear mechanisms and will demonstrate how the careful selection of an appropriate lubricant coupled with a properly targeted lubricant management program can both extend engine and lubricant life particularly where contaminated gas is used.

INTRODUCTION

In any engine regardless of the fuel in use lubricant selection plays a key role in its successful operation. A careful balance of base oil and additive that goes to make up the finished product is very important, for it is the product performance level that will control corrosion, deposits and wear within the engine.

Corrosion can take place on any oil wetted surface within the engine, in the bearings, short of disassembly, corrosive wear cannot be seen, however in the combustion space with the aid of a borescope, the results of a corrosive mechanism can be observed.

In the case of both bearings and bores, lubricant/wear metal analysis coupled with the use of specially balanced lubricants can be a principal means of controlling such corrosion. Good lubricant management and housekeeping plays a key role in ensuring the circulating contaminants are not allowed to build to a level at which the lubricants protective properties can be overwhelmed.

Engines fueled by natural gas and by sewage digester gas have run trouble-free for many years and in many cases with the original wearing components still in service. However modern engines present new problems from those built in the past, they run three times faster and are spark ignited rather than duel fuel operated, however it is reasonable to assume that they will have a similar speed related component life providing

they are correctly lubricated, operated and maintained.

The use of engines to run on landfill gas fuel is relatively new. Only a few years ago, it would have been hard to get a builder guarantee for such applications, however, over recent years it has been proven that landfill gas fuelled engines can be successfully operated and builder guarantees obtained. The other principal change is the electrical generating capacity of the modern engine which is grown from a few kilowatts to over a megawatt in output.

The selection of the correct lubricant coupled with a properly established oil analysis and management program has a critical role to play in the successful operation of all types of spark ignited gas engine where minimum maintenance and optimum engine life are key objectives.

GASES AS FUEL

The internal combustion engine is particularly well suited to the use of gas as a fuel as its homogenaity with the essential combustion air allows for near perfect combustion conditions. However the range in variation in the gases employed as fuels create a variety of widely differing problems because of the physical and chemical characteristics and particularly the gas borne and gaseous contaminents. The combustable components of the fuel gas inevitably range on a scale from light to heavy gases and those which form the heavy end of the spectrum can cause combustion problems. Where methane is a principal gas the level of the heavier gas content should be limited to 5%.

The principal gas used in the gas engine is methane occurring either naturally as extracted from North Sea operation by the gas mains, or as Biogas as a product from the biological degredation either from refuse on landfill sites or by sewage digestion. Both landfill and sewage arising methanes have differing characteristics and contaminants the typical characteristics of the gas are likely to be as follows:

- Methane up to 60% maximum
- Carbon Dioxide up to 40% maximum
- Nitrogen up to 10% maximum

Contaminates Vary, the principal ones being :

- Chlorine-found in landfill and some sewage operations.
- Hydrogen Sulphide-found principally in sewage digestion but also in landfill.

Biogas is principally a mixture of:

- Hydrocarbons
- Inert gases
- Chemically aggressive gases

This mixture will inevitably affect overall performance and may in many cases have an adverse affect on the lubricant charge.

In biogas fueled engines corrosion can be the reaction between the aggressive chemicals and gases with the engine component surfaces. The corrosion mechanism resultant from the cocktail of water, dissolved oxygen, nitrous oxide and general biogas contaminates takes two principal forms

a) the reaction products dissolve in and are removed by the oil and if over-contaminated the oil will carry the dissolved aggressive components to the oil wetted surface within the engine.

b) the reaction products form brittle, abrasive, metal oxides, which break away from affected surfaces.

Both conditions increase lubricant contamination which will have an accelerating affect on the corrosive reaction. This type of wear mechanism results from the acidic components of combustion from contaminated gas which, when combined with water arising from either low temperature operation or when the dew point is reached, combine with chlorine or hydrogen sulphide to attack the following:

Chlorine	Hyrdogen Sulphide	
Aluminium	Copper	
Iron	Iron	(See Fig. 1)

LUBRICANT REQUIREMENTS

The lubricant requirements for liquid hydrocarbon fueled engines are based on reliable, engine driven, oil classifications and specifications. Unfortunately these specifications cannot be applied to gas engines and therefore can only be used as a guide in making a lubricant selection. In selecting the lubricant for the gas engine other specific considerations must be taken into account, these will include:-

- Type of engine - application and duty
- Type of gas and level of contaminates
- Type of lubricant and level of additives with particular reference to the sulphated ash component.
- Type of additive technology to be employed.

The use of a detergent/dispersant engine lubricant is essential in all internal combustion engines as a principal method of controlling wear and maintaining good internal cleanliness. The level of additive treatment must be determined by the likely aggressive nature of fuel contaminants such as sulphur where the content can vary from .5% to over 4%. The greater the level of aggressive contamination of the fuel gas, the greater the quantity of additive addition to the finished lubricant is necessary to control corrosion. Conversely where low levels of fuel contaminates are experienced excessive additive treatment will result in the formation of a deposits which will inevitably cause ignition and valve problems.

One of the performance indicators commonly used to judge the quality of a lubricant is its total base number or TBN. This may or may not be coupled with a measurement of the Sulphated Ash content of the lubricant. In isolation neither of these indicators have much to say about the likely performance in a gas fueled engine. Although there is no ready measurement of the chemical and thermal stability of the additive package and its capability for the prevention of corrosion and formation of deposits these would be true indicators of performance.

The control of deposits is fundamental in ensuring the efficient operation of the gas engine, the formation of deposits will be governed by a number of parameters such as

- Engine design and rating
- Application and operation
- Maintenance
- Type of Gas
- Type of Lubricant

Gas engines are more sensitive to combustion space deposits than their diesel fueled brothers and therefore anything that forms deposits should be avoided.

In developing the discussion of lubricant requirements it is necessary to establish an understanding of alkalinity/detergency/dispersancy/sulphated ash/total base number, which are all inter-linked and are briefly described as follows:

Total Base Number (TBN)

Is a measure of relative alkalinity, the higher the number the greater the alkalinity reserve or neutralising capacity. Unfortunately there is no genuine correlation between TBN and likely neutralisation performance. Therefore on its own the total base number does not provide a reliable indicator of in service performance capability. In used lubricant analysis however the rate of TBN depletion is very useful in measuring the additive's performance in service.

Detergent

The detergent component is the principal additive which determines both total base number and sulphated ash content. It is usually an organometallic compound and is polar in nature. It coats the oil wetted metal surfaces within an engine and prevents the formation of deposits. More importantly, it neutralises acids due to its alkaline nature. Its primary function is to keep the pistons and ring belt clean and free from deposits.

Dispersance

Dispersants are non-metallic, high molecular weight organic compounds and are often referred to as ashless additives. Their main function is to keep unburnt particles of combustion (carbon) in suspension and generally contribute to the cleanliness of the engine.

Sulphated Ash

Is the amount of incombustable material left after burning its lubricant carrier off in a standard oven and crucible test. The level of sulphated ash is not necessarily an indicator of likely deposition in engine combustion spaces, however it is a very useful guide in selecting the correct lubricant for optimum performance, the selection criteria must be to ensure a balance that allows the control of corrosion without the formation of deposits.

Viscosity

Perhaps the most important requirement in any lubricant selection is viscosity (the measure of the oils 'thickness'). The selected viscosity must be one which ensures that there is no metal to metal contact in the loaded zones of the engine at anticipated operating temperatures.

(See Fig. 2)

Viscosity is temperature related and decreases with increasing temperature whilst conversely increasing with decreased temperature. The viscosity requirement of the engine is therefore critical and is established by design which must ensure an adequate oil film is generated at all times.

The very nature of the operation of the internal combustion engine can give rise to both increases and reductions in viscosity, it is therefore essential that the lubricant analysis programme controls viscosity within a fairly narrow band as both increase and reduction in viscosity can be harmful to the engine.

(See Fig. 3)

OIL ANALYSIS

On its own used lubricant analysis is of little value, however, properly applied as part of a planned maintenance programme and supported by periodic physical inspection, it can do much to both prolong the charge life of the lubricant in service, and reduce the incidence of unscheduled maintenance downtime. In installing an oil analysis programme it is therefore essential that clear objectives are established at the outset which will encompass the need for maintenance related actions resultant from the information provided by oil analysis. The information provided by a successful oil analysis programme is well proven but often fails because at inception, engine metallurgies, operating factors, types of gas, operating speeds, output ratings, volume of oil in circulation and consumption of lubricant during service cycle have not been properly assessed.

The first stage of oil analysis must deal with the internal lubricant changes. The following parameters would be checked:

1. **Viscosity** should be checked at both 40 and 100 C, changes will indicate :

   - High insolubles
   - Oxidation
   - Nitration
   - Fuel dilution (Duel fuel engines only)
   - Low load operation
   - Water contamination

   Other analysis areas will indicate the cause of changes in viscosity.

2. **Insolubles** indicates that the oil is thickening from its own internal degredation, or by the ingress of external debris. In a duel fuel engine it could be soot from the pilot fuel or from excessive or prolonged operation on liquid fuel.

3. **Oxidation**, as the name implies, it is a reaction between the lubricant and oxygen, just as rust is reaction between iron and oxygen. Certain contaminates within the engine have a catalytic reaction which can accelerate oxidation resulting in lubricants thickening particularly in high temperature operation forming organic acids which

- increase viscosity
- reduce TBN
- increase TAN

It may also indicate an oil with poor dispersancy.

(See Fig. 4)

4. **Nitration** can occur in all engines but only becomes a problem with spark ignition engines operating just below Lambda 1. It is a combination of nitrogen and oxygen during combustion to form NO2 which causes the oil to thicken forming deposits in the combustion spaces and sludge in the engine, it may also cause filter blocking and shows in oil analysis as :

- increased viscosity
- reduced TBN
- increased TAN

5. **Water** The ingress of water to the gas engine must be kept to an absolute minimum especially where the engine is operated on contaminated biogas. Water forms as part of the combustion process and inevitably will find its way to the sump and where engine operating temperatures are too low it can combine the lubricant charge and combine with the aggressive components of the biogas to form aggressive acids, quite apart from causing rapid degredation of the lubricant charge.

Many of the engine oils developed for the modern liquid hydrocarbon fueled engines cannot be used in the gas engine for when such contamination is likely it can be the very additives themselves that may acclerate corrosion. It is therefore extremely important when selecting a lubricant supplier that they can demonstrate the ability to lubricate a wide range of gas engines from a product line not designed as an offshoot from transport lubricant technology.

WEAR METAL ANALYSIS

Should be used in conjunction with oil condition analysis and be based on engine size, speed and type of fuel gas, it should be operated over set intervals to ensure accurate periodic assessment of the results obtained.

Engine component wear is influenced by two conditions, those components that are load sensitive and those that are speed sensitive, the former is controlled by the gas consumed and the latter by time. To ensure proper interpretation is placed upon the results of wear metals analysis it is important that the analysis results are controlled within tight limits. Corrosion from gas contaminates is unlikely to be evident in mains gas operation however operations utilising Biogas will evidence a more aggressive corrosive condition, this should be taken into account when interpreting the results of wear metal analysis and appropriate allowances be made.

Once a database has been established for any engine operation then the rate of change of wear metal concentrations becomes more important than the assessment of individual results. In all oil analysis, rogue results do occur as a result of poor sampling techniques, however experience tells us that providing no unusual gases or contaminates have been introduced into the engine a check sample will eliminate the need for maintenance action to be taken around such a rogue result. The resultant wear metal trends can be expressed in a simple graphical format and it will be the slope of the graph indicating the rate of change between results which will form the most useful maintenance indicator and show which corrective action will be most appropriate. It is therefore advisable in any oil analysis programme that co-operation between the lubricants supplier the engine builder and the engine operator establishes what is an acceptable and unacceptable level of wear metal contamination. A buffer zone can be established which will reflect the repeatability of the oil analysis results coupled with an application of lubricant knowledge.

(Fig. 5)

A sudden rise in one metallic element may or may not indicate wear, one element from one reading can be very misleading when attempting to determine what is actually happening within the operated engine. Such a reading could indicate a temporary loss of lubricant (resultant perhaps from low oil level) therefore it is more important to react when two or more compatible metallic elements rise and continue to rise, this is a positive indication that wear it is taking place. The only metallic element which may rise in isolation is iron and the increasing presence of iron in the oil analysis without any related metallic elements may indicate that corrosive wear is taking place in the bores and an urgent inspection utilising a borescope would be recommended.

There are a number of different laboratory methods used for determining wear metal content of used lubricant samples, each method has advantages and disadvantages and on balance providing only one method is used as part of a programmed analysis then the results can be used as reasonably true determinants of overall condition. However when establishing a program particularly when a previous oil analysis technique has been in use the first result from the new programme (maybe new laboratory) should be used as a master result and wear metal trends and determinations be established from the master result. The most common wear metal analysis method in use is atomic emmission spectrophotometry which can provide analysis of wear metal particles below ten microns in size. Where it is established that the likely wear characteristic from a large power unit may result in particles in excess of 10 microns then ferrography should be considered as a more appropriate method. A large number of wear metal elements can be tested for but it is always advisable to limit the elements reported to those commonly found in the particular engine operated.

LUBE OIL CONSUMPTION

In addition to oil and wear analysis lubricant consumption can be a useful guide to the engine's general condition, engines with high or above average consumption will tend to accelerate the oil's deterioration from excessive blowby and thus will add to the corrosion problem from contaminated gas at low temperature operation.

Lube oil consumption is proportional to the percentage load, the higher the load the higher the oil consumption and it is for this reason that oil consumption is referred to generally as a percentage of the fuel consumed. Factors influencing consumption will be :

- engine size

- speed and load

- aspiration

- maintenance and operation

- viscosity of lubricant

In the case of viscosity the use of an SAE 40 oil instead of an SAE 30 oil may assist in the reduction of oil consumption whilst improving the firing ring combustion seal. The decision to increase the viscosity of the circulating lubricant will essentially be a decision arrived at by discussion between the lubricant supplier, the operator and the engine manufacturer and will inevitably draw on previous experience with the type of engine operated coupled with the factors already discussed.

LUBRICANT MANAGEMENT PROGRAMME

The key to achieving satisfactory engine operation has already been demonstrated as being a properly set up lubricant analysis programme.

Any such management programme must identify the lubricant condition and its effect on the engine and above all result in actions to be taken prior to the ocurrence of failure. It should set out to show:

- the condition of oil in service

- what changes have taken place and why

- determine an oil change period

- indicate component wear

Once an oil drain period has been established its frequency should be set slightly below the safe limit and should cater for normal changes within the engine and still give adequate protection.

Boroscope

Another valuable aid to the lubricant management programme is the periodic borescopic examination of the combustion spaces. This will show the condition of the following

- valves, including faces, seats and any deposits

- liner for corrosion at top dead centre and in the dead volume of space for carbon or deposit cut-back

- piston crown for deposit build-up and what effect, if any, it is having.

Boroscopic inspection on a regular basis serves to confirm :

- Upper cylinder condition

- Correct lubricant application

- Wear indications from wear metal analysis

and should be seen as essential to a meaningful oil management programme.

TYPICAL GAS ENGINE LUBRICANT SPECIFICATION

Typical is perhaps the wrong word to use in the context of gas engines, experience has shown us that biogas engines operated on sewage farm digester gas just 20 miles apart, have widely differing lubricant needs, thus the below list of parameters for the design of a gas engine oil is of necessity, wide, it does not include certain special operating conditions which from time to time are encountered and must be treated on a one off basis.

Base stocks - The building block of all quality lubricants are inevitably the base stocks which form as much as 98% of the finished product. Here the use of a carefully selected base stock of high chemical and thermal stability is essential to achieve satisfactory gas engine operation. The base stock needs are dictated generally by the higher temperatures of operation experienced in gas engines as opposed to their liquid hydrocarbon fueled sisters which will tolerate a somewhat less fastidious approach.

Sulphated Ash - The sulphated ash content of a lubricant is primarily determined by the metallic compounds arising from the additive package and may include such metals as zinc, calcium, barium, etc. For the gas engine, the sulphated ash content will generally range between 0 and 1% maximum

Total Base Number - A TBN of 6 should be considered a minimum for duel fuel engines but in a spark ignition engine burning clean natural gas a much lower total base number can be tolerated. It must be remembered that, as discussed earlier, TBN is only a guide number and has very little to say about a lubricant's actual performance in service.

Viscosity - Typically the selection of an SAE30 grade will be made when a gas engine has to operate out of doors and be required to start at low ambient temperatures. However, where oil consumption control coupled with indoor operation is a criteria then the use of SAE40 gas engine oils should be considered. Multi-grades are generally not currently used in the gas engine market place although some formulations are undergoing engine manufacturer evaluations at the present time.

SELECTING THE CORRECT PRODUCT FROM ASH AND TBN SPECIFICATION

It is believed that this paper may have already demonstrated that the selection of the appropriate lubricant for a gas engine cannot really be made from sulphated ash or TBN detail, for example note the three following lubricant ash and TBN specifications:

	Oil A	Oil B	Oil C
Sulphated Ash %	2.0	2.0	1.0
TBN	7.0	15	6.5

All three lubricants show that by looking at either ash or TBN for any two products similarities can be drawn. All three indicate specifications of lubricants actually found be lubricating gas engines which were in trouble. In one case severe corrosion was causing engine failures at as little as 2/2,500 hours. In another excessive ash deposits on piston crowns and on valve stems and valves created a need for upper cylinder servicing at 25% of the recommended frequency. It is always difficult for a lubricant manufacturer to cite the cause of the engine problems as being a competitive product but in each case it was in fact true.

The missing element necessary to make proper lubricant selection was the nature and metal type of the additives making up the total base number and sulphated ash content. The incorrect selection was therefore made on the basis of sulphated ash and TBN which, although appropriate in each case for the application contained badly selected additives.

CONTAMINATE LIMITS

The below tabulation should be used as a guide only but is helpful in establising some of the operating thresholds which may require different lubricant considerations.

Chlorine	Below 60 ppm	No problem
	60 to 160 ppm	Caution, special lubricants and filters should be considered.
	160 to 240 ppm	Gas treatment required.
	Above 240 ppm	Severe problems may be encountered
Hydrogen-Sulphide	Below 500 ppm	No problem
	500 to 1000 ppm	Caution, special lubricants and filters should be considered
	Above 1000 ppm	Gas treatment required.

LUBRICANT REQUIREMENT

There is a direct relationship between lubricant, wear metal increase, and oil deterioration in service which shows as :

Increased

- viscosity
- insolubles
- oxidation and nitration
- total acid number
- sulphated ash (from debris ingress)
- wear metals

Decreased

- Total base number
- sulphated ash (from additive depletion)

If we look at ash, for example, where the builder has set strict limits to control combustion space deposits, this can increase in service due to a build up of wear debris, such a buildup may co-incide with the depletion of the detergent additive being neutralised by combustion contaminates and therefore pose no combustion related problem. Such a trend can only be established from a reliable oil monitoring programme, however, if on the other hand the increase is not wear debris but oil additive, then the oil is likely to be at fault either from the incorrect selection of type of product or from a cross mixture with a high detergent lubricant.

LUBRICANT TREATMENT

The treatment of lubricants on site may form part of a lubricant management programme and can take two forms either filtration or purification with the best method depending upon

- Engine size
- Type of gas.

It is well known that an engine's life is determined by the amount of debris ingested. The greatest wear occurs when the particle size of such debris is smaller than the working oil film thickness in the loaded zones (larger particles can't enter the clearances).

The main forms of treatment

- Bypass filters - where around 10% of the oil flow is treated, this method is very effective where the level of contamination is low.

- Full flow filters - where the entire flow is treated before returning the engine.

- Purification (batch or continuous) this method is found mainly on the larger engines due to its costs.

(See Fig. 7)

The chart shows quite clearly that batch purification offers the best overall method of controlling lubricant bourne engine contaminates but also demonstrates that charge life on the smaller engines can be more than adequately extended by the correct application of bypass or full flow filtration.

LUBRICANT FAILURE ANALYSIS

As already demonstrated the gas engine is prone to failure characteristics which are quite unlike those normally experienced in the liquid hydrocarbon fueled engine. Some of its problems are inevitably caused by the aggressive nature of the gas born contaminates arising as the result of the use of biogases either from digester plants or from landfill sites. We have demonstrated thus far that a properly controlled lubricant analysis programme coupled with periodic borescopic inspection and wear metals analysis can be a major contributor in extending the useful charge life of the lubricant whilst identifying potential areas for failure prior to running catastrophe.

There remains just one more area which needs discussion as part of an analysis of the likely causes and remedies of gas engine failure, and that

is the incorrect selection of the lubricant itself. Obviously in the areas already discussed we have identified the need to adopt a lubricant which will satisfactorily cope with the contaminates presented as part of the gaseous fuel but in making the appropriate selection it is often the case that the product itself may give rise to failure characteristics such as

- Exhaust valve burning, from ash built up on valve faces or seats
- Valve sticking through excessive deposits.
- Pre-ignition from piston deposits becoming incandessant
- Spark plug fouling
- Heat exchanger fouling

The correct selection of lubricant is therefore resultant from balancing of all the factors necessary to achieve reasonable operation and it may in some cases be appropriate to select a product which deposits an acceptable level of ash to ensure adequate neutralisation of aggressive gas born compounds and thus the arrest of corrosive wear.

(Fig. 6)

In every case this balance will be struck as a result of the collection of all available data on a site-by-site, engine-by-engine basis, there remains no longer any need to approach the subject of gas engine lubrication with guess work but with certainty and purpose.

CONCLUSION

With energy costs rapidly increasing and the combined heat and power concept now becoming an economic reality, the future of the gas engine looks brighter than ever before in its long history. These engines and their resultant emmissions can without any doubt be classified as environmentally friendly when measured against their petrol and diesel offspring.

However those engines burning contaminated biogas still have to be improved. Lubricants and their correct and properly targeted management have a major part to play in achieving optimum engine life with minimum maintenance down time , it may even be that the lack of such properly targeted programmes may be the limiting factor in successful commercial exploitation of these most desirable energy sources. As for the future we have no doubt that synthetic gas engine oils will have a major part to play both in the extension of lubricant life (particularly in the naturally gas fueled engine) and the control of lube oil related emissions. Our experience with synthetics (polyalphaolefins) suggests an ability not only to comfortably extend lubricant life over the mineral counterpart simply through greater oxidation and thermal ability, but also an enhanced stability to handle circulating contaminates without detrimental effect on the engines operated.

ACKNOWLEDGEMENTS

The authors wish to thank the Mobil Oil Corporation and Mobil Oil Co Ltd along with various builders operators, installers and filter manufacturers for their assistance in compiling this paper.

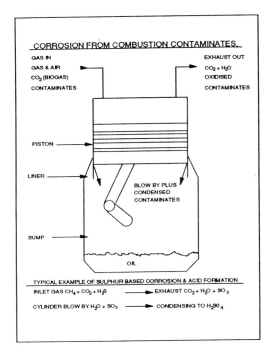

Fig 1

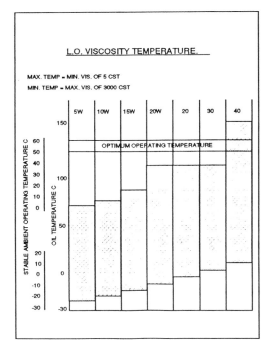

Fig 2

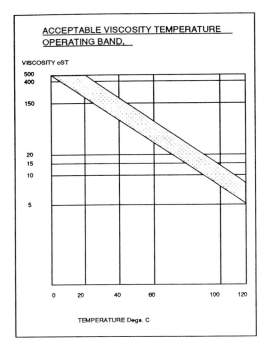

Fig 3

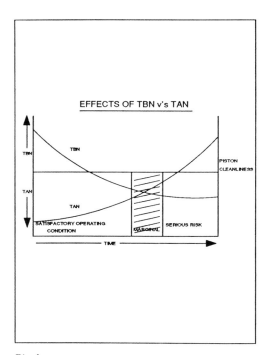

Fig 4

114

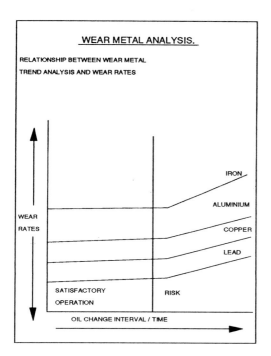

Fig 5

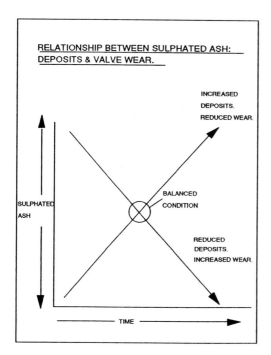

Fig 6

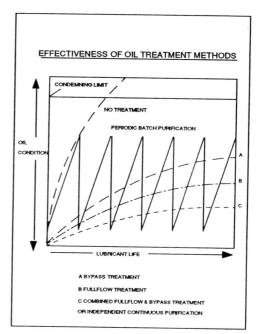

Fig 7

Catalytic exhaust gas cleaning for engines used in co-generation

S BLUMRICH, B ENGLER, W HONNEN and **E KOBERSTEIN**
Department of Physical Chemistry, Degussa AG, D-6450 Hanau, FRG

SYNOPSIS   Three-way catalytic converters are used to purify the exhaust gas of lambda-controlled gas engines. This reduces simultaneously CO, HC and NO_x. Engines with lean operation need a SCR system with ammonia-injection mostly used in combination with an oxidation catalyst, the REDOX system.

1   INTRODUCTION

The reduction in the emission of pollutants, especially from combustion processes, is a main objective of measures which have been taken in order to counteract the impact on the environment which is rapidly continuing to rise.

In addition to the so-called primary measures taken at existing systems (optimization of the course of combustion, conversion to fuels with lower emission levels, etc.), the group of emission-reducing measures also includes retrofitting in the form of secondary measures for the purpose of attaining considerably more pronounced effects in the reduction of emissions.

In view of the so-called "greenhouse effect", a discussion has arisen relating to the efficiency factors of power generation plants. In this regard, Figure 1 compares the energy expenditures for separate generation of electrical power and heat with those of co-generation. The saving in primary energy of 38 % shown in this figure must be equated to a generation of CO_2 reduced by 38 %. This number alone gives an indication of the direction in which the generation of power and heat will develop.

Coupled power and heating plants exist in a wide variety of types and designs. One model which is frequently encountered is the so-called "Combined Heating and Power Plant" with an internal-combustion engine as generator drive, and from which the waste heat streams (exhaust gas, cooling water, lubricating oil) are used for the generation of heating energy. Since, in internal-combustion engines and in gas turbines as well, the amounts of pollutants in the exhaust gas can be reduced by only a small proportion through primary measures, and frequently at the expense of incurring disadvantages in the power area, in very many cases secondary measures are necessary and are also reasonable.

Catalytic exhaust gas purification processes offer almost ideal prerequisites for these cases, and have therefore largely prevailed. Degussa AG, proceeding from the basis of the many years of experience acquired in this field, constantly advances the development of these processes, and tests them for the different cases of application.

2   THE DEFINITION OF THE CATALYST

In chemical textbooks, catalysts are described as follows: Catalysts are materials which increase the speed of a chemical reaction, thereby causing the chemical equilibrium to be established more quickly (Fig. 2).

The catalysts are not consumed in these reactions. In contrast to the temperature, the catalysts do not affect the condition of equilibrium. A catalyst exerts no influence on a system in a condition of equilibrium because it accelerates the forward reaction and the reverse reaction equally.

Through the use of a suitable catalyst, the temperature at which a chemical reaction proceeds at an adequate rate is lowered.

Catalysts are materials of very widely differing character: metals, metallic oxides, non-metallic oxides, bases, acids, but also organic substances as well.

Many catalysts have a specific effect, that is, they are capable of accelerating only certain, very specialized chemical reactions. This specific character of a catalyst prevents undesired secondary reactions from likewise being accelerated (selectivity).

A differentiation is made here between:

Homogeneous catalysis:

The catalyst forms a homogeneous blend together with the reacting materials (gaseous mixture, solution), and

Heterogeneous catalysis:

Here, the reaction medium and the catalyst form different phases. The catalysis itself takes place at a phase interface.

The catalytic purification of exhaust gases is an example of heterogeneous catalysis, as will be described later in greater detail.

3 CONSTRUCTION OF THE EXHAUST GAS PURIFICATION CATALYSTS

From the definition of heterogeneous catalysis, it is immediately possible to recognize the construction of the exhaust gas purification catalysts: while the "reaction medium" is a gas, the "catalyst" is formed of solid matter. The acceleration of the reactions takes place at the surface of the catalyst, resulting in the requirement for a surface that is as large as possible. The possibilities for representing large surfaces in small volumes are not all too varied:

(a) Small shaped bodies (pellets) in spherical form, annular form, tubular form, etc., or
(b) Honeycomb bodies: bodies through which parallel channels pass; the term is clearly derived from the honeycombs found in beekeeping.

Since the honeycomb bodies have largely prevailed for exhaust gas purification catalysts in Europe, we will not present a description of the pellets here.

3.1 Catalysts on ceramic, monolithic supports

The basic structural element of a monolithic catalyst is the ceramic honeycomb body through which exhaust gas channels pass (Fig. 3). For large-scale, commercial use, primarily cordierite has prevailed, which is formed into honeycomb bodies by the extrusion process. Cordierite, a magnesium-aluminum silicate, is used because of its low coefficient of thermal expansion and the associated good resistance to thermal shock. Depending on their application, ceramic monoliths differ from each other with respect to the cell structures, the cell densities, and the external geometrical dimensions. In the case of large, stationary engines, catalysts are used with cell densities between 100 and 400 cpsi (cells per square inch), i.e., on a surface of about 2.5x2.5 cm there are between 100 and 400 square cell channels. The wall thicknesses lie between 0.2 and 0.55 mm.

3.2 Catalysts on metallic, monolithic supports

Metallic monoliths consist of smooth metal sheets alternating with metal sheets which are sinusoidally corrugated. These sheets are either rolled up or are fitted together in a special process to form a so-called S-matrix, and are then brazed at the ends and at the casing (Fig. 4). In these supports, ferritic steels are used which are creep resistant at elevated temperatures.

Cell densities between 50 and 500 cpsi are used. The wall thicknesses here are distinctly lower than in the case of ceramic supports, lying between 0.04 and 0.07 mm.

In comparison with ceramic monoliths, metallic supports are distinguished, for example, by a lower thermal capacity, lower exhaust gas back pressure, larger geometrical surface, and higher thermal conductivity.

3.3 Designs of supports

Both support materials can be processed to cylindrical as well as to rectangular monoliths. Cylindrical elements are preferably used in cases of smaller volumes. Larger reactors are equipped with rectangular monoliths (Fig. 5).

3.4 Casings

(a) Ceramic supports:

Round monolithic catalysts are "packed" in sheet metal casings (Fig. 6). A wire mesh mat or a thermally stable fiber mat is fitted between the ceramic material and the sheet metal for two purposes: in the first place to protect the ceramic material, and in the second place to absorb the different thermal expansions.

Rectangular monoliths are built into casings which can be matched to the particular space conditions. Here also, thermally stable fiber mats are used for purposes of the stabilization.

(b) Metallic supports:

The external sheet metal enclosing the round metallic supports can be used as casing, so that no additional expenditure is necessary here. In case of need, only a flange must be welded on for the mounting. Rectangular monoliths can be welded directly together with cones to form a reactor.

4 COATING TECHNOLOGY

4.1 Precious metal catalysts

The catalytically inactive support monolith is coated in a complicated production process with an oxidic intermediate layer which consists mainly of alumina. As a result, the high surface area mentioned earlier, which can be as large as 20,000 m^2 per liter of catalyst volume, is attained. The precious metals in finely divided form are then precipitated in the pore system of this intermediate layer (Fig. 7).

In order to improve the sintering stability and the catalytic activity, elements from the series of the rare earth metals are additionally incorporated into the oxide intermediate layer.

Figure 8 shows a view through some coated channels.

4.2 NO_x reduction catalysts

The manufacture of NO_x reduction catalysts for the SCR process is similar to the process just described. However, the composition of the coating is made up of other materials: depending on the case of application, catalysts can be manufactured with coatings based on zeolite or with the active materials vanadium pentoxide and tungsten oxide. Fundamentally, the possibility also exists of using catalysts fully made of catalytic material - these catalysts consist entirely of catalytically active material, and they are used in large power plants because of their stability with respect to abrasion. For applications with an engine drive, they have not been able to prevail because of the large structural volume required.

In addition to the catalyst formulation, i.e., the chemical composition, other parameters also play an essential role for the suitability of the catalysts, for example the distribution of the individual components, the pore structures which influence the mass transfer, the surface roughness, etc. Suitable combinations of these characteristic values make it possible to match the catalysts to the various exhaust gas purification concepts.

5 THREE-WAY CATALYSTS

In the three-way catalyst, the actual catalytically active components are the precious metals platinum and rhodium, occasionally also palladium.

By means of the three-way catalyst the three pollutants nitrogen oxides, carbon monoxide, and hydrocarbons can be removed simultaneously in one single catalyst (Fig. 9). In this catalyst, the nitrogen oxides are reduced to nitrogen through separation of the oxygen bonded to the nitrogen, and at the same time the hydrocarbons as well as the carbon monoxide are oxidized to carbon dioxide and water. The most important chemical reactions are summarized in Table 1.

In order to ensure that the pollutants are converted as effectively as possible at the catalyst to the environmentally neutral substances nitrogen, water, and carbon dioxide, it is necessary to feed an almost stoichiometrically composed exhaust gas containing oxidizable and reducible components into the catalyst. If there is too much oxygen in the exhaust gas, the reduction of the nitrogen oxides will take place to only a slight degree because of the oxygen requirement that is lacking, while if the exhaust gas contains too little oxygen, the hydrocarbons and the carbon monoxide can be only incompletely oxidized (Fig. 10).

In order to derive the stoichiometrically composed exhaust gas, so-called lambda regulating systems are used, the construction and function of which will be explained later.

6 OXIDIZING CATALYSTS

In the oxidizing catalyst as well, platinum and palladium, more rarely rhodium, are used as catalytically active components.

With the oxidizing catalyst, hydrocarbons and carbon monoxide can be removed from the exhaust gas (Fig. 11). Both materials are oxidized to carbon dioxide and water with the oxygen contained in the exhaust gas. The prerequisite for a conversion that is as complete as possible is sufficient availability of oxygen at an adequately high temperature. The oxygen content in the exhaust gas should lie above 2% by vol.; an example of the temperature dependence is represented in Figure 12.

The most important chemical reactions are summarized in Table 1 in Equations 1 to 3.

The collective term "hydrocarbons" includes many different compounds which are oxidized at very widely differing rates. Saturated hydrocarbons, especially methane, react the most slowly, while unsaturated organic compounds containing oxygen as well as polycyclic aromatic hydrocarbons are in the middle of the field. Carbon monoxide and hydrogen react much more quickly.

7 NOx REDUCTION CATALYSTS

There is no such thing as the NO_x reduction catalyst! Instead, a multitude of catalysts exists which are suitable for the conversion of nitrogen oxides to nitrogen under certain conditions. In the cases of stoichiometrically composed exhaust gases, this reaction proceeds as described earlier in the case of the three-way catalyst. With exhaust gases which contain excessive oxygen, the SCR process (SCR = Selective Catalytic Reduction) presents itself as a possibility; in this process, ammonia must be added to the exhaust gas as reducing agent for the nitrogen oxides (Fig. 13).

The mechanism of this catalytic reduction is very complicated. On the catalyst surface, various reactions proceed side by side at the reaction centers. For example, oxygen also intervenes in the reaction processes, NO and NO_2 are equally reduced. Other exhaust gas components such as sulfur dioxide and water likewise influence the reaction. In Table 2, the most important equations are listed. As end products, nitrogen and water are developed. However, if the process is conducted unfavorably, nitrogen oxides can also develop from the ammonia (Table 2, below).

The reaction mechanism will not be explained in greater detail at this point. It should only be shown that, at a ratio of ammonia to nitrogen oxides of 0.7 to 1.0, a reduction in the nitrogen oxides of 70 to 90% is highly successful. The limiting factor for the conversion - at a given catalyst size - is the permissible ammonia "slip", i.e., the amount of ammonia which passes unconsumed through the catalyst. For technical as well as for environmental protection reasons, this amount must be kept as small as possible. Figure 14 shows the nitrogen oxide residual emission as a function of the amount of ammonia injected. The rising ammonia slip with an increasing excess of the reducing agent can be clearly recognized.

Which catalyst is used for which case of application depends on the particular requirements: conversion required, temperature, space conditions, permissible pressure loss, type of the engine (gas or diesel engine, two-stroke or four-stroke cycle engine, etc.).

8 DIESEL PARTICULATE FILTERS

A technique for the purification of exhaust gases that is encountered more rarely with large engines is the diesel particulate filtration by means of ceramic diesel particulate filters.

The monolithic diesel particulate filter (Fig. 15) consists of parallel channels which are closed alternately at one end or the other. The channel walls are produced from a porous ceramic material to which a catalyst is applied at the inlet end. During the flow of the exhaust gas through the channel walls, the particulates are held back and are ignited by the catalyst at about 550°C. The soot is then burned off, thus resulting in disposal by a method which leaves no residue. In the outlet channels, an oxidizing coating is applied which oxidizes hydrocarbons and carbon monoxide to carbon dioxide and water. After the soot has been burned off, the filter is loaded again.

The limits of this technology lie in the composition of the particulates: if primarily combustible soot is involved, this technology can be used with good success. If, however, the principal component of the soot is an incombustible oil ash, irreversible clogging of the diesel particulate filter occurs very quickly.

9 APPLICATIONS AND EXPERIENCE

Catalysts can be used for the purification of exhaust gases at the following types of engine (Fig. 16):

Gas, spark-ignition engines (4-stroke cycle) in stoichiometric operation;
Gas, spark-ignition engines (4-stroke cycle) in over-stoichiometric operation;
Gas, spark-ignition engines (4-stroke cycle) in lean operation;
Gas, spark-ignition engines (2-stroke cycle) in over-stoichiometric operation;
Dual-fuel engines;
Diesel engines.

The spark-ignition, gas engines, which in terms of their numbers are by far the most frequently encountered, have a typical field of emission characteristics (Fig. 17): in the rich operating range (lambda < 1), high CO, HC, and NO_x emissions are generated, in the over-stoichiometric range (lambda between 1.0 and about 1.5), low CO and HC emissions, but very high NO_x emissions are observed, and in the lean range, NO_x emissions values are found which lie under the limit values of the so-called "TA-Luft" (the German law which regulates the purity of the air), but values of CO and HC which are already rising again.

The limit values according to the TA-Luft are represented in Table 3. Uniform limit values for carbon monoxide, hydrocarbons which do not contain methane, etc. contrast with differing limit values for the nitrogen oxides. Consideration must be given to the so-called "Dynamizing clause", which states that all possibilities must be exhausted to reduce the NO_x emissions further through measures taken at the engine or by other measures corresponding to the "State of the art".

9.1 Three-way catalyst

The three-way catalyst, due to its specific operating conditions already described, can be used reasonably only with engines which are equipped with a mixture regulation system, the so-called lambda regulation (Fig. 18). The lambda sensor installed in the exhaust gas pipe before the catalyst measures the residual oxygen content in the exhaust gas, and feeds this value as control signal to the gas-air mixer. This system must ensure that the engine is operated at a combustion air ratio, lambda, slightly under 1 (Fig. 19). Here is where the correct exhaust gas composition lies for the function of the three-way catalyst. According to control measurements carried out regularly by Degussa in a combined heating and power plant, the waste heat from which is used to heat an indoor swimming pool, NO_x, CO, and HC conversions were measured at stationary gas engines which correspond to those measurements made when the plant was put into initial operation. At the point in time when the measurements were made, the catalysts had operated for 22,000 and 19,200 hours respectively.

9.2 Oxidizing catalyst

The removal of hydrocarbons and carbon monoxide by means of an oxidizing catalyst is the simplest method for purifying exhaust gas because the oxidizing catalyst has a very wide working range. It is therefore largely independent of the exhaust gas composition, and does not presuppose any engine regulation measures.

Oxidizing catalysts are used in large numbers in so-called "lean-burn" engines. In Figure 20, this is represented using the example of a mixture-supercharged, lean-burn engine. These engines also require a regulation system so that they can be operated reliably within the limit values of the TA-Luft. Here also, the term "lambda window" is used (Fig. 21). If an oxidizing catalyst is installed, the control range can be expanded far to the right, toward even leaner operating modes.

Lean-burn engines equipped with oxidizing catalysts attain CO emission values which as a rule lie far below those which are permitted by the TA-Luft.

9.3 REDOX process

In the cases of all engines which operate over-stoichiometrically, NO_x reduction in the exhaust gases is not possible with a three-way catalyst. Here, the NO_x reduction must be carried out by means of the SCR process already described, with the addition of ammonia via SCR catalysts. Usually, this takes place according to the REDOX process, a combined process, during the implementation of which an oxidizing catalyst is installed downstream from the NO_x reduction catalyst (Fig. 22). The exhaust gas generated by the engine must, if necessary, be cooled through a heat exchanger to the catalyst operating temperature. Before entry into the reactor, the required amount of ammonia is mixed as homogeneously as possible with the flue gas. When flowing through the NO_x reduction catalyst, the selective catalytic reduction, which has already been mentioned, takes place. In the oxidizing catalyst which follows, primarily hydrocarbons and carbon monoxide are oxidized to carbon dioxide and water. A processing advantage of this arrangement consists of the fact that ammonia that has passed through the SCR catalyst is likewise oxidized. As a result, the danger of an ammonia slip, that is, release of ammonia into the atmosphere, is precluded. The amount of ammonia supplied can be controlled, for example, by means of a family of NO_x load characteristic curves of the engine or through exhaust gas analyses. When families of NO_x load characteristic curves are used, a quick reaction of the ammonia metering system to changes in the NO_x in the exhaust gas caused by load changes is ensured.

9.4 Double-bed process

The purpose of the double-bed process is to attain high conversion rates of the pollutants CO, HC, and NO_x to N_2, CO_2, and H_2O during a long period of operation. The difference with respect to the three-way process consists of the decoupling of the reducing and oxidizing processes. Similar to the case with the three-way process, the engine must be equipped with a lambda regulation system. Figure 23 shows schematically the construction of a pilot system: The exhaust gas initially flows through a three-way catalyst, in which the reduction of the nitrogen oxides takes place according to the three-way principle. Then air is mixed into the exhaust gas, and at the oxidizing catalyst which follows, the hydrocarbons which have not yet been converted, as well as the carbon monoxide, are oxidized.

The advantage of this arrangement is that the engine adjustment must take only the best possible reduction of the nitrogen oxides into consideration. Removal of the other two pollutants takes place independent of the engine adjustment.

First measurement results after 1000 hours of operation at the pilot plant show the effectiveness of the concept; especially, it was possible to attain a distinct reduction in the hydrocarbon emissions (Fig. 24).

9.5 Diesel particulate filter and NO_x reduction catalyst

In the cases of diesel engines, the particulate emission presents additional environmental problems. The definition of the particulates includes carbon particulates (soot) together with hydrocarbons and inorganic compounds adsorbed on them.

In conjunction with a pilot experiment, the exhaust gas purification system shown in Figure 25 was studied: In the first stage, the exhaust gas coming from the diesel engine flows through a diesel particulate filter in which up to 50% of the particulates emitted are collected. The rate of collection here depends greatly on the porosity of the ceramic material. After passing through a heat exchanger, the NO_x reduction is carried out according to the SCR process.

Figure 26 shows the effectiveness of the catalytic particulate collection during the course of about 2300 hours of operation. The rate of reduction lies at about 55%. The pressure loss of the filter passed through a flat maximum. During a check of the system after 2300 operating hours, it was possible to prove excellent operating properties.

As an alternative to the diesel particulate filter, an oxidizing catalyst can be used. The effect of the particulate reduction is attained through the fact that above all the hydrocarbons adsorbed at the carbon particulates (soot) are oxidized.

10 AGEING AND REDUCTION IN ACTIVITY OF CATALYSTS

10.1 Ageing

Similar to every part of a system or plant, the catalyst is also subject to a certain ageing. This is synonymous with a decrease in the activity. The main cause for this is a very slow, temperature-dependent sintering of the catalytically active coating (Fig. 27). The sintering processes which take place at elevated temperatures have the result that the specific surface of the material becomes smaller. This transformation process is associated with the operating time and the temperature of the exhaust gas which has been generated. The higher the exhaust gas temperature, the more quickly the transformation proceeds.

10.2 Dangers of poisoning for catalysts

The increasing utilization of sewage and disposal dump gases as fuels for internal-combustion engines has led to some problems with respect to exhaust gas purification. These combustible gases often contain heavy metals as well as arsenic, halogen, sulfur, and/or phosphorus compounds which are then found again in the exhaust gas from the engine (Fig. 28). In a downstream catalyst, some of these pollutants can very easily collect, and reduce the activity of the catalyst by a blockage of the active center.

Experience has shown that three-way catalysts react more sensitively to catalytic poisons whereby,

in contrast, the oxidizing catalyst is much more robust.

10.3 Coating of ash

A deactivation of the catalyst can also develop through a clogging of the catalyst channels. In these cases, a fine, white coating can be found on the inlet end of the catalyst. Analyses of these coatings have shown typical lubricating oil constituents such as calcium, phosphorus, and zinc as main components. It is therefore assumed that these coatings are residues from the combustion of the lubricating oil. Since these particles occur as a mechanical covering of the catalyst inlet surface, the activity of the catalysts can generally be reestablished through removal of the coatings.

11 SUMMARY

To summarize, the exhaust gas purification methods are represented once again in Figure 29.

The three-way catalyst with lambda sensor regulation can be used for the purification of exhaust gas from gas, spark-ignition engines in the lambda-1 operating mode. At the same time, carbon monoxide, hydrocarbons, and nitrogen oxides in the exhaust gas can be reduced.

In the cases of engines with exhaust gases which contain excessive oxygen, it is possible to reduce the carbon monoxide and the hydrocarbons through the use of an oxidizing catalyst. The nitrogen oxides can be reduced by selective catalytic reduction, for example in the REDOX process, with ammonia.

Table 1

Chemische Reaktionen bei der katalytischen Abgasreinigung

Chemical reactions during the catalytic cleaning of exhaust gas

$$C_MH_N + (M + N/4)O_2 \rightarrow MCO_2 + N/2\,H_2O \tag{1}$$

$$CO + 1/2\,O_2 \rightarrow CO_2 \tag{2}$$

$$H_2 + 1/2\,O_2 \rightarrow H_2O \tag{3}$$

$$CO + NO \rightarrow 1/2\,N_2 + CO_2 \tag{4}$$

$$C_MH_N + 2(M + N/4)NO \rightarrow (M + N/4)N_2 + N/2\,H_2O + MCO_2 \tag{5}$$

$$H_2 + NO \rightarrow 1/2\,N_2 + H_2O \tag{6}$$

Table 2

$$(1) \quad 4\,NO + 4\,NH_3 + O_2 \longrightarrow 4\,N_2 + 6\,H_2O$$

$$(2) \quad 2\,NO_2 + 4\,NH_3 + O_2 \longrightarrow 3\,N_2 + 6\,H_2O$$

$$(3) \quad 6\,NO + 4\,NH_3 \longrightarrow 5\,N_2 + 6\,H_2O$$

$$(4) \quad 6\,NO_2 + 8\,NH_3 \longrightarrow 7\,N_2 + 12\,H_2O$$

$$(5) \quad 4\,NH_3 + 7\,O_2 \longrightarrow 4\,NO_2 + 6\,H_2O$$

Table 3

Emission limits for Stationary Engines in West-Germany (TA-Luft)

Type of Engine	Fuel	Total Fuel Heat Cont. of Plant	Nitrog oxides (NO$_x$)	Carbon-monoxid (CO)	Hydrocarbons without Methan (NMHC)	Soot
Gas Engine (4-cycle)	Natural Gas	>1MW$_{th}$	0,5	0,65	0,15	–
Gas Engine (2-cycle)	Natural Gas	>1MW$_{th}$	0,8	0,65	0,15	–
Dual-fuel Engine and Diesel Engine	Natural Gas and or Diesel	1–3MW$_{th}$	4,0	0,65	0,15	0,13
		>3MW$_{th}$	2,0	0,65	0,15	0,13

The limits are listed in g m^3 i.N. and are referred to 5 Vol % of oxygen in Exhaust Gas.

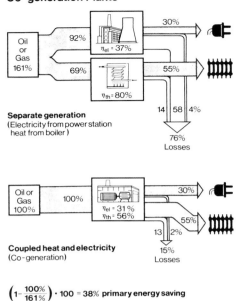

Energy Saving by using Co-generation Plants

$$\left(1-\frac{100\%}{161\%}\right) \cdot 100 = 38\% \text{ primary energy saving}$$

Fig 1

Definition of a catalyst:

A catalyst is a material
which increases the rate
of a chemical reaction,
this causes a faster chemical
equilibration.

Fig 2

Keramische Monolithträger
Ceramic substrates

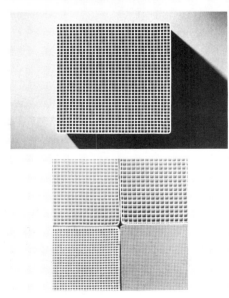

Fig 3

Metallischer Monolithträger
Metallic substrates

Fig 4

Bauformen von Katalysatoren
Catalyst design

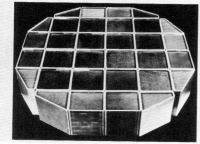

Fig 5

Gehäuse
Casing

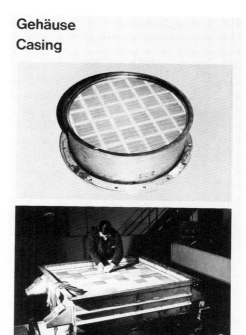

Fig 6

Querschnitt durch einen Katalysator
Cross-section of a catalyst

Fig 8

Aufbau der Beschichtung
Structure of Coating

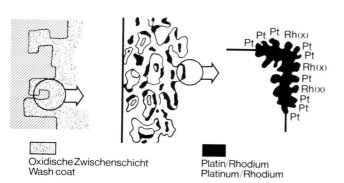

Oxidische Zwischenschicht
Wash coat

Platin/Rhodium
Platinum/Rhodium

Fig 7

Schadstoffminderung durch Dreiwegkatalysator
Pollutant reduction by means of three-way catalyst

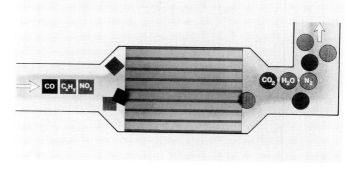

Fig 9

Konvertierungsverhalten eines Dreiwegkatalysators
Conversion-ratio of a three-way catalyst

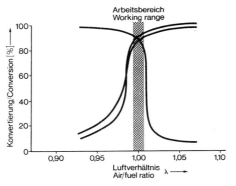

Fig 10

Schadstoffminderung durch katalytische Oxidation
Pollutant reduction by means of catalytic oxidation

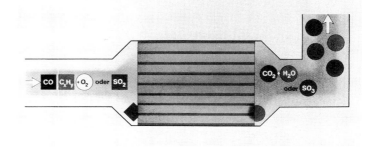

Fig 11

Konvertierung von Kohlenoxid und Kohlenwasserstoffen an Oxidationskatalysatoren
Conversion of carbon monoxide and hydrocarbons

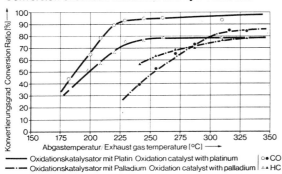

Fig 12

Selektive katalytische Reduktion mit Ammoniak
Selective catalytic reduction with ammonia

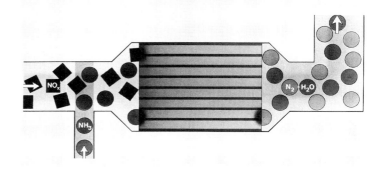

Fig 13

NOₓ-Reduktion und NH₃-Schlupf als Funktion der NH₃-Einspeisung
NOₓ-Reduction and Ammonia Slip as a Function of Ammonia Injection

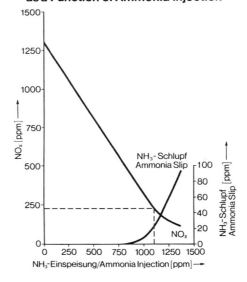

Fig 14

Aufbau und Wirkungsweise eines Dieselrußfilters
Design and Operation of a Diesel Particle Filter

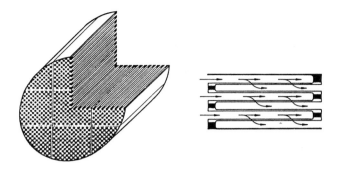

Fig 15

Engine Types used in Co-generation Plants

○ Gas Engines (4-cycle) with stochiometric combustion (Air/Fuel Ratio 16.00:1)

○ Gas Engines (4-cycle) with slightly lean combustion (Air/Fuel Ratio 16.07:1 – 17.00:1)

○ Gas Engines (4-cycle) with lean combustion (Air/Fuel Ratio 18.00:1)

○ Gas Engines (2-cycle)

○ Diesel Engines

○ Dual-fuel Engines

Fig 16

Emissionen eines Otto-Gasmotors als Funktion von Lambda
Exhaust emissions from a gas engine as a function of air/fuel ratio

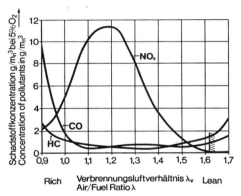

Fig 17

Gas-Ottomotor mit Dreiwegkatalysator
Gas Engine with Three-way Catalyst

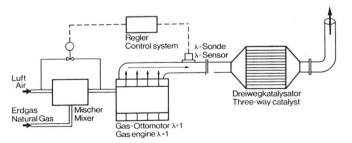

Fig 18

Schadstoffkonvertierung eines Dreiwegkatalysators als Funktion von Lambda
Three-way catalyst conversion ratio

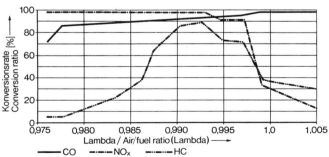

Fig 19

Katalytische Abgasreinigung für gemischaufgeladene Magermotoren
Catalytic Exhaust Gas Cleaning for Lean-burn Engines

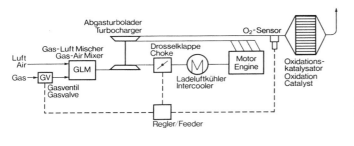

Fig 20

Magermotoren Lambda-Fenster im Magerbereich
Lean-burn Engines Lambda-Window

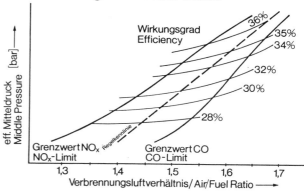

Fig 21

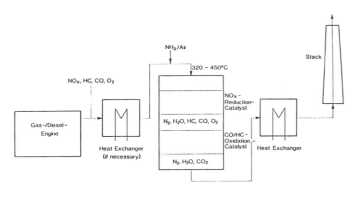

The REDOX Process for Exhaust Gas Purification
in Cogeneration Plants Using Stationary Gas- or Diesel – Engines

Fig 22

Doppelbettverfahren/Dual Bed operation

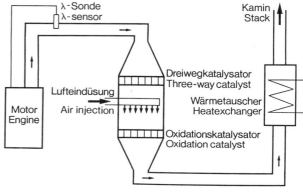

Fig 23

Doppelbettverfahren/Dual Bed Operation

Konvertierung des Oxidationskatalysators
Conversion rate of oxidation catalyst

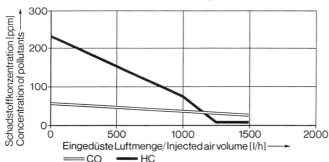

Fig 24

Katalytische Abgasreinigung bei Dieselmotoren
Rußabscheidung und NOₓ-Reduzierung

Exhaust Gas purification for Diesel Engines

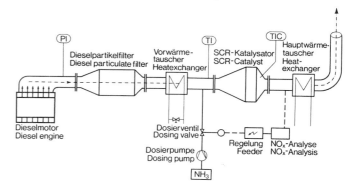

Fig 25

Dieselpartikelfilter/Diesel particulate filter
Rußabscheidung und Druckverlust als Funktion der Betriebszeit
Soot separation and loss of pressure

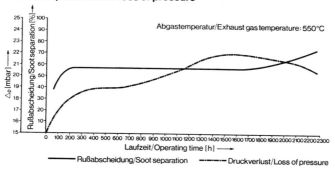

Fig 26

Alterung der Katalysatoren
Ageing of Catalysts

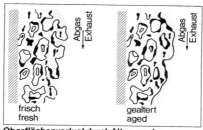

Oberflächenverlust durch Alterung der oxidischen Zwischenschicht

Ageing of »washcoat«

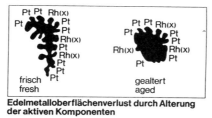

Edelmetalloberflächenverlust durch Alterung der aktiven Komponenten

Ageing of precious metals

Fig 27

Desaktivierung von Katalysatoren
Deactivation of catalysts

1. Vergiftungsgefahren für Katalysatoren	**Poisoning of catalysts**
○ Schwermetalle	○ heavy metals (Pb)
○ Halogene	○ halogens (HF, HCL)
○ Arsenverbindungen	○ arsenic compounds
○ Schwefelverbindungen	○ sulphur compounds
○ Phosphorverbindungen	○ phosphorus compounds
○ Zinkverbindungen	○ zinc compounds
2. Aschebelegung	**Ash deposits**

Fig 28

Catalytic Emission Control System for Stationary Engines

Type of Engine	Emission Control System
Gas engines (4-cycle) with stochiometric combustion (A/F-Ratio: 16.00:1)	Three-way Catalyst
Gas engines (4-cycle) with slightly lean combustion (A/F-Ratio: 16.07:1 – 17.00:1)	Selective Catalytic Reduction
Gas engines (4-cycle) with lean combustion (A/F-Ratio: 18.00:1)	Oxidation Catalyst
Dual-fuel engines	Selective Catalytic Reduction
Gas engines (2-cycle)	Oxidation Catalyst
Diesel engines	Selective Catalytic Reduction Oxidation Catalyst Soot Filter

Fig 29

Use of selective catalytic reduction for reduction of NO$_x$ in gas engine exhaust

P MORSING, MSc and **K SØNDERGAARD**, MSc
Haldor Topsøe A/S, Lyngby, Denmark

SYNOPSIS

The SCR process is today the most widely used process for the reduction of NOx in flue gases from coal, oil or gas fired boilers. The process can also be used for achieving a high reduction of NOx in the exhaust gas from gas fired engines. A general introduction to the SCR process is given and the incorporation as well as the control of the process on a gas engine is shown. The influence of different gas engine and exhaust gas related parameters on the design and the economics of the SCR unit are discussed.

1. INTRODUCTION

The use of combined power and heat production has potentially economical as well as environmental advantages, as the utilization of the energy source can be more than doubled compared to power production alone. Large combined heat and power plants are in operation in densely populated areas in large cities, where the heat can be distributed to many consumers without prohibitive large investments and heat losses in the distribution net. The interest for obtaining the benefits of this co-generation in small communities or for large single consumers has led to an increased interest in small plants based on gas turbines and gas engines for economical reasons, but also for environmental reasons.

Gas is a clean fuel with respect to SO$_2$ and particle emissions. However, the uncontrolled combustion of gas in reciprocating engines results in a high NOx emission, as the NOx formation during combustion is mainly controlled by the high temperature in the combustion zone in the cylinders of the engine. The NOx emission from medium to high output (2 - 50 MW$_e$) diesel gas engines is typically 800 - 3000 mg/MJ energy input (about 450 - 1700 ppmv at 15% O$_2$ in the exhaust gas) while the NOx emission, for comparison, from pulverised coal or oil fired boilers is typically 125 - 500 mg/MJ input energy (about 200 - 700 ppmv at 6% O$_2$ in the flue gas). In fact, as it appears from Table 1, the NOx emission per fuel energy input from diesel engines is about the highest from any of the normally used fuel combustors.

As a consequence of the high NOx emission from gas engines, a substitution of power and heat produced on the basis of oil and coal fired boilers with power and heat produced in gas engines without any NOx reducing measures may therefore result in an increase in the total NOx emission from the power and heat production sector.

In the following sections a description of the availability of NOx reducing methods for gas engines is presented with special emphasis on the SCR (Selective Catalytic Reduction) process.

2. NOX EMISSION STANDARDS AND REDUCTION METHODS

Today only a few countries have NOx emission standards on gas engines. However, with the growing concern for the environment, more countries will most likely follow in the near future. In Table 2 a collection of existing and proposed emission standards for stationary engines in various countries is shown. The emission standards in the Table are expressed in ppmv at 15% O$_2$ in the exhaust gas, although the actual standard may have a different basis.

The proposed Danish standard is expected to become effective in the beginning of 1990, and similar to the Dutch standard, which incorporates a correction for the engine efficiency (shaft energy/fuel energy), the Danish standard incorporates a correction for the electrical efficiency (electrical energy/fuel energy) of the engine. Thus a credit is given for a high efficiency in the energy conversion from gas to electricity. For small engines, with an electrical efficiency of appr. 30%, the emission standard is 120 ppm at 15% O$_2$, whereas for large engines, with an electrical efficiency of appr. 45%, the emission standard is 180 ppm at 15% O$_2$.

For the main part of the shown standards, it is necessary to substantially reduce the engine NOx emission in comparison with what is achieved by ordinary gas engines without any emission control. For some of the shown standards, a NOx reduction of more than 90 - 95% is necessary for diesel engines with high, uncontrolled NOx emissions.

The NOx reduction methods are, in parallel with the nomenclature used for boilers, divided into two categories:

1) Primary methods.
2) Secondary methods

The primary methods are defined as changes in the engine construction or operation that lead to a reduced NOx formation during the combustion process. The secondary methods are defined as measures, which reduce the NOx emission by removing NOx from the exhaust gas.

Of the primary methods, the lean-burn combustion technolgy is the most efficient, as NOx emissions below 100 ppm (at 15% O_2) can be achieved.

By this technique the combustion takes place with a high excess of air corresponding to a stoichiometric ratio of about 1.6. However, it is only applicable to engines with spark ignition and today it is offered only for engines with an electric output of up to appr. 1 MW_e. The lean-burn technique leads to increased CO and HC emissions, and in some cases it may be necessary to install an additional catalytic combustion unit for removal of these components. The technique is not applicable to diesel engines (dual fuel engines), which cover the medium to large engine sizes (2 - 50 MW_e).

Exhaust gas recirculation, water injection and lowering of the maximum combustion pressure are other potential primary methods. These have, however, mostly been tested on oil powered diesel engines, and generally they can only offer a low NOx reduction which is insufficient to meet modern NOx emission standards.

Of the secondary methods, only the catalytic methods are used for gas engines. Three-way catalysts, which are used for NOx, CO and HC reduction from automobile exhausts can also be used for spark ignited gas engines operating without excess air (at a stoichiometric ratio of 1.0). This method is, however, today only offered for smaller engines with electric outputs of less than about 1 MW_e.

For medium to high output gas engines, which all operate with excess air (both spark ignited and diesel engines), the SCR process will therefore often be the only possibility to obtain a high NOx reduction to meet modern NOx emission standards.

3. THE SCR PROCESS

The SCR process is the most widely applied process for the reduction of NOx in exhaust gases. The process is particularly applied in Japan and in the Federal Republic of Germany, where SCR units for use in power plants, with a total production capacity of more than 75 000 MW_e (corresponding to the production from about two hundred large power plant units), have been built or ordered. The SCR process can be used for flue gases from either coal, oil or gas fired boilers or for exhaust gases from gas turbines and diesel engines (gas or oil powered). With the use of the SCR process the emission of nitrogen oxides can be reduced by 90% or more.

The reduction of the nitrogen oxides takes place by injecting ammonia into the exhaust gas at an exhaust gas temperature of 300 - 400°C and letting this gas mixture flow through a catalyst where the nitrogen oxides, which primarily consist of NO and small amounts of NO_2, are converted according to the following reaction schemes:

$$4NO + 4NH_3 + O_2 = 4N_2 + 6H_2O$$

$$6NO_2 + 8NH_3 = 7N_2 + 12H_2O$$

As can be seen from the above schemes, the conversion of the nitrogen oxides does not create any secondary pollution components. The products formed are only nitrogen and water vapour, which are already present in the atmosphere in large amounts.

The degree of NOx removal depends on the amount of ammonia added (expressed by the NH_3/NOx ratio). At high NH_3/NOx ratios, a high degree of NOx removal can be obtained, but at the same time, the amount of unused ammonia (called the NH_3-slip) in the cleaned exhaust gas will increase.

It is normally desirable that the concentration of unused ammonia in the cleaned exhaust gas is as low as possible, because when the exhaust gas cools in the downstream boiler or heat exchanger, the ammonia may react with any SO_3 in the exhaust gas and lead to a fouling of the heating surfaces by ammonium sulphates. For cleaning sulphur oxide free exhaust gases, as in the case of spark ignited gas engines or exhaust gases with very low sulphur oxide contents, as in the case of diesel engines, the amount of unused ammonia has no or only very little impact on operation.

4. PROCESS LAYOUT AND SIZING OF THE UNIT

In Figure 1 the SCR process is shown schematically. The main components of the SCR process consist of a reactor with catalyst and an ammonia addition system.

The ammonia can either be liquid, water-free ammonia under pressure or it can be a 25% aqueous ammonia solution at atmospheric pressure. The ammonia or ammonia water is evaporated in an electrically or steam heated evaporator and is subsequently diluted with air before the mixture is injected into the exhaust gas channel.

The injection of the ammonia/air mixture into the exhaust gas channel takes place through a system of nozzles in order to achieve a homogeneous mixing of the ammonia with the exhaust gas. A static mixer may be placed in the exhaust gas channel to further improve mixing. It is important to obtain a homogeneous NH_3/NOx ratio in the exhaust gas in order to achieve an efficient operation of the SCR process and minimize the NH_3-slip from the SCR reactor.

The SCR reactor contains one or more layers of catalyst. The catalyst volume, and consequently the size of the reactor, depend on the activity of the catalyst, the desired degree of NOx reduction, the NOx concentratration, the exhaust gas pressure and the acceptable NH_3-slip. The amount of catalyst can be expressed by the term space velocity (abbreviated NHSV), which is defined as the number of normal cubic metres of exhaust gas which are treated per cubic metre of catalyst per hour.

In Figure 2 an example of how the NOx reduction and the NH_3-slip vary with the NH_3/NOx ratio for two different catalyst volumes (NHSV) is shown.

As can be seen in Figure 2, both the NOx reduction and the NH_3-slip increase with an increasing NH_3/NOx ratio. It can also be seen that it is possible to obtain the same NOx reduction (for example 80%) by using only half the catalyst volume (NHSV = 10 000 Nm^3/h instead of 5000 Nm^3/h) just by increasing the NH_3/NOx ratio by a few per cent.

At the same time, however, the NH_3-slip goes up considerably. The maximum acceptable NH_3-slip has thus a great influence on the required amount of catalyst.

Normally SCR units are designed for a maximum NH_3-slip of 5 - 10 ppmv during the lifetime of the catalyst, when the exhaust gas contains sulphur oxides, in order to minimize the previously mentioned deposits of ammonium sulphate in the downstream heat exchanger or boiler. For exhaust gases with no sulphur oxides or only very low concentrations of these, the SCR unit can be designed for a higher NH_3-slip. Figure 2 also illustrates that if a higher degree of NOx reduction is desired, it is necessary to increase the catalyst volume in order to keep a low NH_3-slip at the same time. At a higher degree of NOx conversion the demand for accuracy of the control of the ammonia addition (NH_3/NOx ratio) is increased, because if the NH_3/NOx ratio becomes larger than 1.0, the NH_3-slip will increase drastically due to the break through of unused ammonia.

The catalyst normally has a monolithic structure, which means that it consists of blocks of catalyst with a large number of parallel channels, the walls of which are catalytically active. In this way deposits of dust on the catalyst are minimized and the pressure drop over the catalyst is also kept low. The channel diameter has an influence on the volume activity of the catalyst and on the dust deposition.

With the use of a catalyst with a relatively small channel diameter for example, the necessary catalyst volume will be reduced but at the same time the pressure drop over the catalyst will increase as well as the risk of dust deposition on the catalyst.

The catalyst is manufactured in a number of different channel diameters, so that the channel diameter can be optimized after study of the dust content of the exhaust gas, the characteristica of the dust and the allowable pressure drop across the SCR reactor. The exhaust gas from gas engines cannot be considered as being particle free as it in many cases (as for diesel engines) can contain 50 -100 mg/Nm^3 of particles. The relatively high particle concentration is due to the combustion of lubricating oil.

5. PROCESS CONTROL

The control of the ammonia addition is shown in Figure 3. The control is normally performed by means of a small process computer, which calculates the necessary amount of ammonia to be injected on the basis of continuous measurements of the amount of gas (or of the load on the engine) and on the NOx concentration in the exhaust gas together with a given NH_3/NOx set point. This control can also be combined with a feed-back control on the basis of a continuous measurement of the NOx concentration in the cleaned exhaust gas in order to secure a more accurate control of the ammonia addition. This is particularly necessary for high degrees of NOx reduction. In its simplest form the control of the amount of ammonia can be based on the engine load and on a memorized curve giving the dependency of the NOx concentration in the exhaust gas on the load of the engine. This simple control scheme can only be used at relatively low degrees of NOx reduction (50 - 70%).

On diesel engines, where the exhaust gas always contains minor amounts of sulphur oxides from the combustion of ignition oil, it is necessary to interrupt the addition of ammonia if the exhaust gas temperature drops below 300 - 310°C at low engine loads. Otherwise the catalyst will be deactivated by deposits of ammonium sulphates. For spark ignited engines, where the exhaust gas is sulphur oxide free, the addition of ammonia can continue at considerably lower temperatures (down to appr. 250°C).

6. INTRODUCTION OF SCR REACTORS FOR GAS ENGINE POWER PLANTS

The optimum operating temperature of the catalyst (300 - 400°C) dictates the position of the SCR reactor in the exhaust gas path after the gas engine. In Figure 4 three different positions of the reactor are shown. The reactor can be positioned between the turbocharger and the waste heat boiler, as shown in case a. This is the typical position for a four-stroke diesel engine. The reactor can also be positioned before the turbo-charger, as shown in case b. This is the typical position for a two-stroke diesel engine. For certain engines the exhaust gas temperature can be too high upstream of the waste heat boiler, in which case the reactor must be positioned as an integrated part of the waste heat boiler as shown in case c. In this case the exhaust gas temperature is lowered to appr. 350°C in the superheater upstream of the reactor.

The necessary catalyst volume, and consequently also the reactor size, will be smaller when placed before the turbocharger due to the higher pressure (typically 3 bar abs), but positioning the SCR reactor here may, in certain cases, be more difficult.

In order to avoid condensation of water on the catalyst during a start-up of the gas engine after a long standstill, which has allowed the reactor to cool down, the reactor must be equipped with either an electric or a steam pre-heating system.

An SCR unit on a gas engine co-generation plant will typically be operated at a constant load and only have few daily load changes. In contrast to operation of an SCR unit for a peak load power plant or for a ferry engine, where relatively frequent and fast load changes are experienced, there will be no high demands to the speed of controlling the ammonia addition.

7. CATALYST CONSUMPTION

The catalyst deactivates slowly during operation, mainly due to thermal ageing and physical blocking of the catalyst surface by dust. After a certain period of operation on the original catalyst charge, the performance of the catalyst is no longer adequate.

This period of operation is somewhat misleadingly called the lifetime of the catalyst. It is typically two to five years, but depends on the initial catalyst volume and operating conditions. What is more important, from an economical point of view, is the catalyst consumption over the economical evaluation period of the plant, for example 15 years. As illustrated below, the catalyst consumption can differ considerably from the catalyst consumption defined by the reciprocal of the catalyst lifetime, if an appropriate replacement schedule is used. To illustrate the effect of different replacement schedules, an example of the change in the catalyst activity over a 15 year operating period for three different replacement schedules is shown in Figure 5.

The catalyst activity shown, k, is the activity relative to the total fresh activity, k°, of the initial charge. The performance of the catalyst is adequate until the activity has reached a value of $k/k°$ designated as the design point. It is assumed that a replacement or addition of catalyst can only take place at a yearly outage placed at intervals of 1 year.

The reactor contains three catalyst beds, initially filled with catalyst in all three cases. However, in the last case, c, the reactor is provided with an extra empty catalyst bed for later installation of additional catalyst.

The simplest of the replacement schedules is shown in case a, where all the initially filled catalyst layers are replaced by new catalyst, whenever the total catalyst activity reaches the design point. In this way the catalyst consumption is equal to 12 catalyst layers the first 15 years, including the initial charge of three layers. This corresponds to an average catalyst consumption of 0.80 layers per year.

This replacement schedule results in a catalyst consumption which should be equal to the initial catalyst volume devided by the catalyst lifetime, which in the example is 4 years. This results in a catalyst consumption of 0.75 layers per year (= 3 layers/4 years). The reason for the small difference is due to the fact that in case a, the catalyst activity is still above the design point after 15 years. Had the evaluation period been 16 years in case a, then the catalyst activity would have dropped to the design point and the catalyst consumption would also have been 0.75 layers per year (= 12 layers/16 years).

A better replacement schedule is shown in case b, where only one catalyst layer is replaced by new catalyst whenever the total catalyst activity reaches the design point.

This replacement schedule allows further use of the catalyst activity in the used catalyst, which is left in the reactor. By using a sequential replacement, all three layers will eventually be replaced and the cycle is then repeated. In this way the catalyst consumption is 9 catalyst layers the first 15 years, including the initial charge of 3 layers. This corresponds to an average catalyst consumption of 0.60 layers per year or a 25% reduction compared to case a.

Today reactors are normally designed with an extra empty catalyst bed. This results in a further improvement of the replacement schedule, as shown in case c. When the design point is reached the first time, an additional catalyst layer is installed in the reactor on the empty catalyst bed. When the design point is reached the second time, a sequential replacement of the initial catalyst charge starts, as in case b. In this way the catalyst consumption is 7 catalyst layers the first 15 years including the initial charge of 3 layers. The average consumption is now reduced to 0.47 layers per year or a reduction of 42% compared to case a.

8. ECONOMICAL ASPECTS

The investment and the operating costs for an SCR unit for NOx reduction in exhaust gases from gas engines depend primarilly on the capacity of the unit, the NOx concentration, the desired degree of NOx reduction and the yearly number of operating hours of the engine.

In Table 3 some examples of investment and operating costs for an SCR unit placed after the turbocharger on a 4 MW$_e$ and a 10 MW$_e$ engine are shown. The unit removes 80% of the NOx content by an assumed NOx inlet concentration of 800 ppmv. The operating time is 4000 equivalent full load hours.

The ammonia source is pressurized liquid ammonia at a price of £ 200/ton. We have not assumed any operating personnel for the SCR unit.

The investment in the SCR reactor (exclusive of catalyst) and ammonia feed and control system including erection, is calculated at £ 250 000 and £ 350 000 for a 4 MW$_e$ and a 10 MW$_e$ engine respectively. When assuming a 15 year depreciation period and an efficient interest of 6% in the calculation of capital costs, the yearly operating costs including interest and depreciation of the investment, become £ 49 800 and £ 88 700 respectively.

If the cost of operation of the SCR unit is put in proportion to the electricity production alone, and not in proportion to the total energy output from the plant, the total operating costs correspond to 0.31 p/kWh and 0.22 p/kWh respectively.

The use of 25% aqueous ammonia instead of pressurized liquid ammonia typically increases the cost of ammonia. The increase is very much site specific, as it depends on the local availability of the two ammonia sources and on the shipping distance from the supplier. Because of the appr. 4 times greater shipping volume of the aqueous ammonia, the cost of this ammonia source is more sensitive to the shipping distance.

9. THE EXPERIENCE OF HALDOR TOPSØE A/S WITH THE SCR PROCESS

The Topsøe deNOx catalyst is produced in standard elements with a height of 0.5 metres and a cross section of 0.5 metres by 0.5 metres or 0.5 metres by 0.25 metres. These standard elements can be assembled in large catalyst modules with larger cross section areas and greater heights than the standard element, depending on the individual application. The catalyst may also be supplied with different channel openings (hydraulic diameters), depending on the dust concentration in the exhaust gas.

The Topsøe deNOx catalyst has been successfully demonstrated in power plant installations and on a stationary diesel engine.

The catalyst has been successfully tested for 4500 hours placed before the turbocharger on a stationary two-stroke diesel engine running on heavy fuel oil (1) and 6000 hours on a 10 000 Nm^3/h SNOX demonstration unit (simultaneous SOx and NOx removal) in a coal fired power plant. Furthermore the catalyst has been tested for 3000 hours on a 10 000 Nm^3/h high dust SCR demonstration unit also in a coal fired power plant.

Haldor Topsøe A/S has recently received orders for the supply of a 310 MW$_e$ and two 30 MW$_e$ SNOX units as well as for two SCR units for installation before the turbocharger on two 8 MW$_e$ two-stroke MAN B&W marine diesel engines. The first of these units was commissioned in December 1989.

Also a small SCR unit has been sold for a 700 kW$_e$ four-stroke gas diesel engine to be started up in February 1990. This SCR unit is a pilot and demonstration unit, where a recently developed high temperature catalyst for operation at 400 - 500°C will be tested. A sketch of the plant is shown in Figure 6. The high temperature catalyst is placed upstream of the superheater in the boiler section, which in this case has a horizontal gas path. It is also possible to demonstrate operation with the standard catalyst placed in the boiler section at a temperature of appr. 350°C, in which case the high temperature catalyst will be removed. The ammonia source is aqueous ammonia and the mixing into the exhaust gas will be done by means of a static mixer. The plant will be operated by remote control from a regional dispatcher.

REFERENCES

(1)   P. Schoubye, K. Pedersen, P. Sunn Pedersen, O. Grøne and O. Fanøe. Reduction of NOx Emissions from Diesel Engines. CIMAC Conference, Warsaw, 1987.

Table 1

Typical NO$_x$ Emissions from Uncontrolled Combustion

Combustion Process:	Typical NO$_x$ Emission		
	ppmv NO$_x$ (% O$_2$)		mg NO$_x$/MJ
Gas fired boilers	50-500	(1)	25-250
Fuel oil fired boilers	200-600	(3)	125-350
Coal dust fired boilers*	400-700	(6)	300-500
Gasoline Spark ign. engines**	1000-4000	(0)	500-2000
Gas (stat.) Spark ign. engines**	300-1000	(5)	200-800
Diesel 4-stroke, oil, gas	600-1400	(13)	800-1900
Diesel 2-stroke, oil, gas	1000-1700	(15)	2100-3100
Gas turbines, oil, gas	100-250	(15)	200-450

1 ppmv NO$_x$ (x% O$_2$), dry gas $\approx \dfrac{a}{21-x}$ mg NO$_x$/MJ (NO$_x$ as NO$_2$)

a = 11.1 for coal and oil. a = 10.3 for gas.

O$_2$ concentration on a dry gas and volume basis.

ppmv: ppm on volume basis.

* Dry bottom boilers. Molten slag boilers about 2 times higher emission.
** Spark ign.: spark ignition.

Table 3

Investment and Operating Costs for an SCR Unit
in English Pounds

	Engine Power	
	4 MWe	10 MWe
Investment:	250 000	350 000
Yearly costs		
Capital costs (10% of investment):	25 000	35 000
Catalyst costs:	8 400	21 000
Ammonia costs:	8 900	22 200
Maintenance (3% of investment):	7 500	10 500
Total yearly costs	49 800	88 700

Exchange rate used: 1 English £ = 11 Danish kr.

Table 2

NO$_x$ Emission Standards for Stationary Engines

Country	Engine Rating	NO$_x$ limits (ppmv at 15% O$_2$, dry gas)
Federal Republic of Germany	Diesel:	
	3 MW$_{th}$ $\geq$ P > 1 MW$_{th}$	730
	P > 3 MW$_{th}$	365
	Otto (4-stroke):	
	P > 1 MW$_{th}$	90
	Otto (2-stroke):	
	P > 1 MW$_{th}$	145
Holland*	Diesel:	
	P $\leq$ 50 kW$_{th}$	Approx. 700 ($\eta_{eff}/30$)
	P > 50 kW$_{th}$	Approx. 235 ($\eta_{eff}/30$)
	Otto:	
	P $\leq$ 50 kW$_{th}$	Approx. 465 ($\eta_{eff}/30$)
	P > 50 kW$_{th}$	Approx. 160 ($\eta_{eff}/30$)
Denmark (Proposal)	50 MW$_e$ > P > 120 kW$_e$	120 ($\eta_e/30$)
Sweden * (Guidelines)		Approx. 60-115
South Korea		88

P: Engine input power η_{eff} : Effective engine efficiency

ppmv: ppm on volume basis η_e : Electrical efficiency of engine

* Standard expressed in mg/MJ energy input

   (conversion used 100 mg/MJ = 58 ppmv at 15% O$_2$, dry gas).

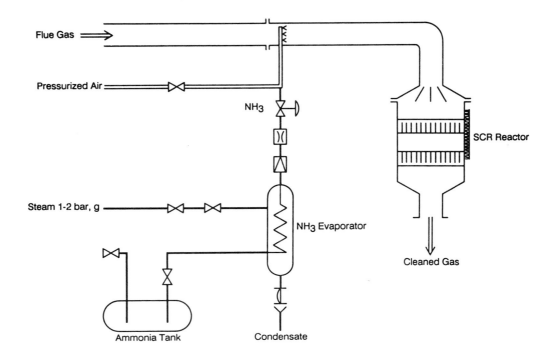

Fig 1   The SCR process

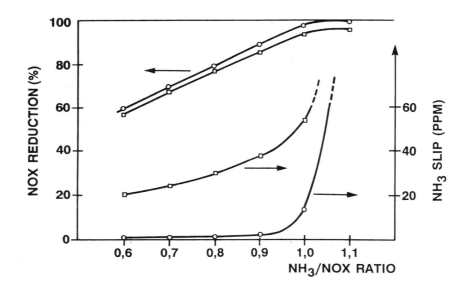

○ : NHSV = 5.000 Nm^3/m^{3}h

□ : NHSV = 10.000 Nm^3/m^{3}h

Fig 2   NO$_x$ reduction and NH$_3$-slip versus NH$_3$/NO$_x$ ratio

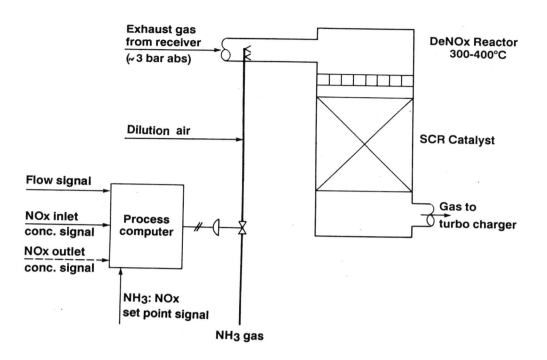

Fig 3   Control of ammonia addition

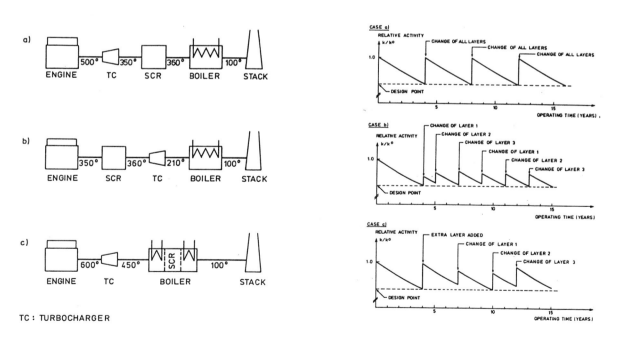

Fig 4   Position of SCR reactor in gas engine plants

Fig 5   Catalyst replacement schedules

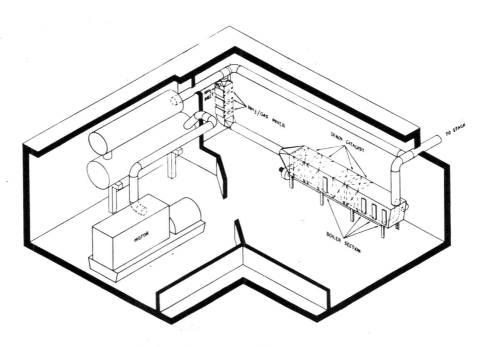

Fig 6   SCR unit for 700 kW gas engine

Emission control for gas engines. Reduction of nitrogen oxides, hydrocarbon, and carbon monoxide levels

G W BLOOMFIELD and **R GUNTRUM**
Johnson Matthey Frankfurt, FRG

SUMMARY:

Exhaust emissions from natural gas or L.P.G. fuelled stationary engines are of a similar composition to those from automobiles. However different operating regimes can produce markedly different exhaust compositions requiring a range of after treatment technologies to lower pollutant levels.

Experience with NOx, Hydrocarbon and Carbon Monoxide removal from various engine applications has shown that three major options are needed.

Work in the F.R.G. has shown the effectiveness of automotive type catalytic convertors when applied to gas engines under certain operating conditions.

BACKGROUND AND INTRODUCTION:

The catalytic control of atmospheric pollutants from automobiles has been firmly established in the U.S.A. and Japan, and is increasingly being adopted by European and other industrial countries.

Exhaust levels of Nitrogen oxides, Hydrocarbons and Carbon Monoxide are reduced by controlling the air:fuel ratio of the engine and by simultaneous conversion of all three pollutants in either a dual bed catalytic converter or in a single bed "three way" catalytic converter.

For so called "three way" catalytic conversion to occur, at the high efficiency levels required, only a narrow window of operation is available, this being dependant on the pollutant and Oxygen levels in the exhaust gas which in turn is governed by the air:fuel ratio setting of the engine.

Emissions from stationary sources, such as power station boilers, stationary internal combustion engines and gas turbines, account for over half of the man-made hydrocarbons, carbon monoxide and oxides of nitrogen reaching the atmosphere. This has prompted strict legislative limits controlling such emissions, both in the U.S.A., West Germany and elsewhere.

For stationery engines combustion is normally carried out using a large excess of Oxygen, in order to maximize fuel efficiency, thus precluding a direct application of automotive "three way" catalyst technology.

However, experience with gas fuelled (spark ignition) engines in the F.R.G. has shown that automotive type catalysts can be successfully applied to stationary engines operating at a stoichiometric air:fuel ratio or an equivalence ratio of Lambda = 1.

EXHAUST CHARACTERISTICS AND CONTROL STRATEGIES:

Various methods for the control of engine exhaust air pollution are available, ranging from engine engineering, through operational changes to aftertreatment

Fig. 1 shows some of these.

All to one extent or another will reduce certain aspects of exhaust pollution, however this reduction of pollution normally incurs a penalty of some type i.e. reduced power or increased specific fuel consumption.

Catalytic aftertreatment systems offer significant improvements in emission levels with the lowest impact on engine performance.

To varying degrees, all engines produce exhausts containing the following components:

Reducing Agents:
Carbon Monoxide
Hydrogen
Hydrocarbons

Oxidizing Agents:
Oxygen
Nitrogen Oxides

Inerts:
Nitrogen
Carbon Dioxide

RICH BURN ENGINES:

With rich burn engines, such as most naturally aspirated engines, the relative amounts of reducing and oxidizing agents are similar and depend on air:fuel ratio. By careful control of the engines air:fuel ratio the exhaust gas composition can be balanced, so that over a suitable catalyst, the naturally occurring reducing agents will react with the oxidizing agents present to form inert compounds i.e.

$$CO + O_2 \longrightarrow CO_2$$
$$H_2 + O_2 \longrightarrow H_2O$$
$$NO_x + CO \longrightarrow CO_2 + N_2$$
$$NO_x + H_2 \longrightarrow H_2O + N_2$$
$$HC + O_2 \longrightarrow CO_2 + H_2O$$

In effect this class of operation should be considered as Lambda = 1, as over rich or very rich engine operation presents additional problems i.e. excessive hydrocarbon and carbon monoxide production and possible formation of ammonia over the catalyst in very rich operating conditions;

$$3H_2 + 2NO \longrightarrow 2NH_3 + CO_2$$

EXHAUST POLLUTION CONTROL ALTERNATIVES

CATALYTIC SYSTEMS:	DE-NOx FOR RICH BURN SYSTEMS SCR FOR LEAN BURN SYSTEMS OXIDATION FOR C0 AND HC REMOVAL
EXHAUST GAS RECIRCULATION:	POSSIBLE REDUCTION OF NOx - UNRESOLVED APPLICATION PROBLEMS TO DATE
WATER INJECTION:	REDUCTION OF NOx FORMATION BY LOWERING OF CYLINDER TEMPERATURES - INJECTED AS FUEL EMULSION.
RETARD INJECTION TIMING:	DEFINITE REDUCTION IN NOx - PENALTIES INCREASED SPECIFIC FUEL CONSUMPTION, C0 AND HC OUTPUT.

Fig 1

The fine balance of exhaust components required for three way catalytic conversion to be effective cannot be achieved by simply presetting the engine carburation system rich. Variations of air:fuel ratio can be caused by changes in local conditions such as air temperature, humidity, fuel pressure, fuel composition and equipment fluctuations.

Here a further import of automotive technology must be utilized, an automotive type oxygen sensor is used to monitor the air:fuel ratio of the engine (by sensing exhaust 02 levels). Adjustments are made by adding or removing small amounts of "trimming" fuel to the intake manifold of the engine through a motorized control valve. Control of the process is managed by a small dedicated microprocessor.

LEAN BURN ENGINES:

Lean burn engines as typified by diesel, two-cycle and many turbocharged gas engines, have the same exhaust gas components but in different proportions. Oxygen levels of 5 - 15% are not uncommon. In these conditions the balance of oxidizing to reducing compounds is distorted, the presence of a catalyst will cause oxidation of hydrocarbons and carbon monoxide to carbon dioxide and water vapour but will do nothing for the oxides of nitrogen.

It is necessary to introduce a reducing agent that will react with the N0x whilst ignoring the O_2. Ammonia is the only agent that will achieve this selective reduction, as it does not normally occur in the exhaust it must be added from an external supply. The level of ammonia required is approximately equal to the level of N0x present.

LEAN BURN LOW N0x

Where the lean burn engine produces a low N0x level, i.e. below local legislative limits, then an oxidation catalyst alone is required to treat the hydrocarbons and carbon monoxide present. This will not effect the small amount of NOx present but will convert the HCs and C0 to carbon dioxide and water.

In summary the three catalyst options are; **(Fig 2)**

For lean mix/high N0x exhaust situations:- typically over 700kw, Selective Catalytic Reduction (SCR), requiring the introduction of a reducing agent - ammonia.
Temperature range for operation; 350 - 420 deg C.
Conversion rates; up to 95% NOx and C0.

For rich mix (or Lambda = 1) engine operation:- typically 100 - 1000kw operation, "three way" catalyst systems (Non SCR, or DeN0x).
Temperature range for operation; 450 - 680 deg C.
Conversion rates; 90 - 95% NOx, C0, and HCs.

For lean mix/low N0x exhaust:- typically 200 - 10000kw, oxidation catalyst alone (HONEYCAT) required to remove C0 and Hydrocarbons.
Temperature range for operation; 250 - 550 deg C.
Conversion rates; 90 - 95% C0 and HC.

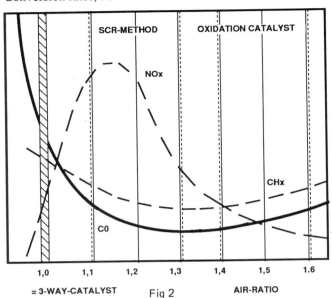

Fig 2

EXPERIENCE IN F.R.G

Legislative requirements for engine exhaust emissions in the Federal Republic of Germany, as outlined in the TA Luft regulations, provide a requirement for control of NOx, C0 and HCs. A number of companies operating gas engine generation sets in the F.R.G. have applied catalyst aftertreatment in order to achieve the required emission limits.

Fig. 3 shows the legislative limits for engine emissions for F.R.G.

LIMITING VALUES OF CO NOx EMISSIONS ACCORDING TO T.A. LUFT		
PROCESS	**POLLUTANT LEVEL**	**NOTES**
COMBUSTION ENGINES Emissions referenced to 5% O2 - dry.	CO 0.65 g/cum NOx 2.0 g/cum NOx 4.0 g/cum	Compression engines 3Mw or more. Compression engines less than 3Mw.
GAS TURBINES Emissions referenced to 5% O2 - dry.	CO 0.10 g/cum NOx 0.30 g/cum NOx 0.35 g/cum	At 60000 cum/hr or more. At 60000 cum/hr or less.

Fig 3

In order to achieve these limits a catalyst system must be designed to suit the specific engine or application. Performance characteristics of the engine are obviously required in order to determine the amount and type of catalyst such as, exhaust gas flow, gas temperature and composition, the type and consumption rate of lubricating oil, water content of the exhaust gas, and criteria such as maximum allowable back pressures and space constraints are also required.

The catalyst composition and support must be optimized to provide the required conversion rates whilst creating a low backpressure on the engine and maintaining a high durability despite mechanical and thermal shock.

DeN0x CATALYST DESIGN

The above combination of criteria for a catalyst system have over a period of time led to an optimal design for the Johnson Matthey DeN0x ("three way catalyst") system.

Considerations of system back pressure, resistance of thermal and mechanical shock and good corrosion resistance have led to the adoption of a honeycomb like support structure manufactured from Armco 18 SR stainless steel. This is constructed into circular monolith blocks of up to 30 inch diameter onto which the active catalyst materials are fixed. This type of support offers minimal backpressure whilst presenting a high surface area to the gas stream. The facility to produce single large diameter catalyst elements from stainless steel avoids the requirement to manufacture the catalyst element from ceramic blocks (most common form of catalyst substratum) which can allow gas bypass between joints and may not be as resilient to mechanical damage.

Fig. 4 shows the low pressure drop characteristics of this type of construction.

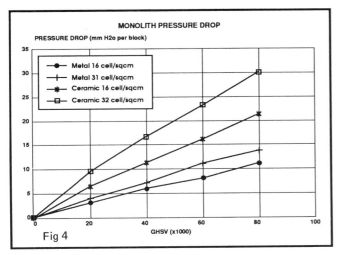

Fig 4

The catalyst materials cannot be fixed directly to the support monolith therefore an intermediate "washcoat" layer is first applied. This porous layer increases the overall surface area of the support and forms a key for the precious metal catalyst materials. The washcoat is an alumina composition, primarily Al_2O_3 with various other oxides. Again selection of the composition of the washcoat is determined by the requirements for high catalyst performance and durability, for the washcoat this is primarily resistance to temperature effects. **Fig. 5** shows the stability of washcoat.

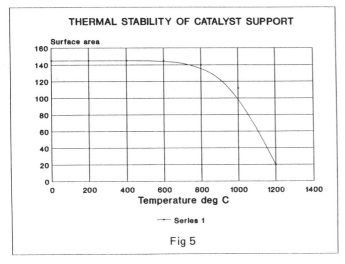

Fig 5

The active catalyst materials comprise a mixture of platinum group matals, such as platinum, rhodium and palladium. The exact composition of this mixture has been formulated to achieve high conversion rates of N0x, C0 and HCs and longevity of operation. Metal loading rates are determined by the level of activity required, a comparison of the activity of different loadings is shown in **Fig. 6**.

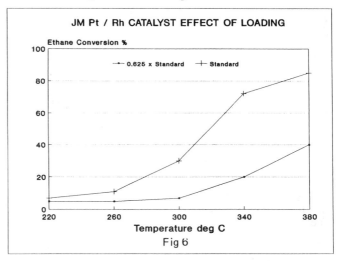

Fig 6

The major considerations in most cases is the catalyst life, optimizing catalyst composition for high activity is essential, but this activity must continue for extended periods. Typical life expectancy for an automotive catalyst would be around 120000 km or 2000 hrs, for stationary engines a potential for operation of over 8000 hrs a year exists, a typical catalyst endurance of 15000 hrs was sought (equivalent to 1000000km for a car).

As can be seen from **Figs 7 - 11** high conversion rates can be maintained over such periods with minimal deactivation.

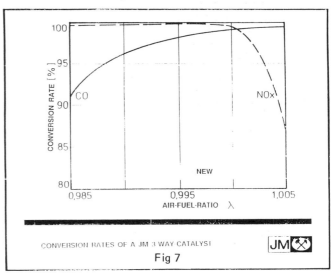

Fig 7

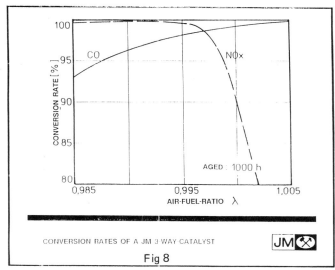

CONVERSION RATES OF A JM 3-WAY-CATALYST

Fig 8

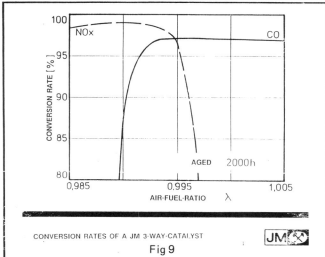

CONVERSION RATES OF A JM 3-WAY-CATALYST

Fig 9

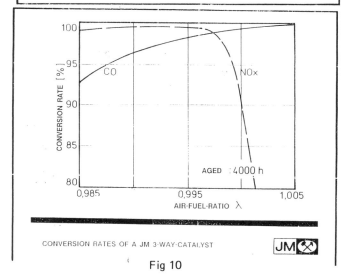

CONVERSION RATES OF A JM 3-WAY-CATALYST

Fig 10

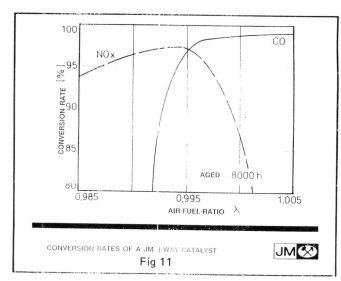

CONVERSION RATES OF A JM 3-WAY-CATALYST

Fig 11

The mechanics of catalyst deactivation over time are three fold, thermal deactivation, poisoning and masking. The first two are not normally reversible as they involve either migration of the catalyst from the substratum or agglomeration which reduces available catalyst sites (thermal) and chemical attack of the catalyst material (poisoning). Masking is retrievable as this involves blockage of the catalyst sites or washcoat pores, which can normally be cleaned off.

In normal engine operation exhaust temperatures are not high enough to cause thermal deactivation problems, from modeling, tests and existing plants it has been found that catalyst life expectance of over 20000 hrs can be expected at an operation temp of 480 deg C.

Poisoning can occur if catalyst attacking compounds enter the fuel or lubricating oil i.e. Lead, Phosphor, Sodium, Arsenic and certain other compounds. Again in normal engine operation this should not occur, although care must be taken to assess fuel and lubricating oil composition to ensure significant levels of the above are not present.

Masking is the normal mechanism of catalyst deactivation for this type of application, by either ash or dust i.e. lubricating oil ash, carbonatious material, scale or non organic combustion products. Several washing procedures are available depending on the nature and severity of the fouling, from compressed air/water/steam for dust and ash to mild detergents for organics or silica through to caustic solutions for persistent and certain inorganic agents.

Fig.12 shows that whilst a drop off of activity will be noticed after extended operating periods the effect on actual conversion rates is small and that the above using the means, recovery to almost the original activity level of the catalyst can be achieved.

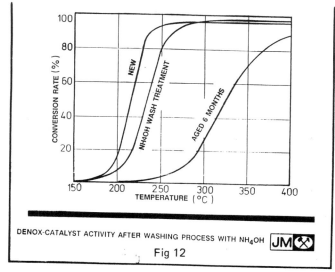

DENOX-CATALYST ACTIVITY AFTER WASHING PROCESS WITH NH₄OH

Fig 12

SYSTEM DESIGN

The overall design of the system is depicted in **Fig. 13.** Location of the catalyst module is determined by exhaust temperatures as catalyst operation is temperature sensitive in terms of conversion efficiency and time to begin operating.

One or more Oxygen sensors are installed prior to the catalyst to provide the necessary exhaust state feedback to the fuel control equipment. Although catalyst life can be extended beyond 15000 hrs, the oxygen sensors have a limited life of around 2000 hrs.

The catalyst is housed in a corrosion resistant stainless steel shell, engineered for even distribution of exhaust gases to achieve peak performance. The module is normally centre flanged to facilitate installation, and removal for inspection or maintenance.

For any given application the amount of catalyst used will be determined, as mentioned previously, from consideration of gas flows, temperatures etc. A variation of catalyst volume for the same application has definate effects on catalyst performance. This is shown in **Fig. 14** which depicts performance against variation of gas hourly space velocity, the design basis for catalyst volume.

ACKNOWLEDGEMENTS

I should like to thank R. Guntrum formally of Johnson Matthey Frankfurt for his input on operating experience in F.R.G.. Also S. Nanos, H. J. Jung, E. R. Becker, C. E. Falletta, A. F. D'Alessandro, L. R. Thomas for material from previous papers on various aspects and applications of catalyst aftertreatments systems.

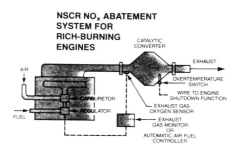

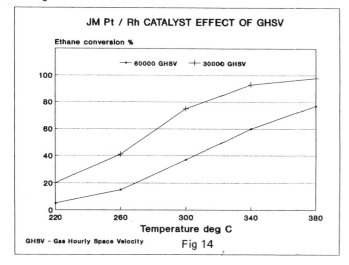

Fig 13

Fig 14

In summation, the Johnson Matthey DeN0x Catalyst has shown that automotive style "three way" or dual bed catalyst systems can be successfully applied to gas engines operating at Lambda = 1 or slightly rich, and can achieve the requirements for pollutant removal to legislative limits over a worthwhile operating life.